Marine Tidal and Wave Energy Converters

Marine Tidal and Wave Energy Converters

Technologies, Conversions, Grid Interface, Fault Detection, and Fault-Tolerant Control

Special Issue Editors

Mohamed Benbouzid
Yassine Amirat
Elhoussin Elbouchikhi

MDPI • Basel • Beijing • Wuhan • Barcelona • Belgrade • Manchester • Tokyo • Cluj • Tianjin

Special Issue Editors

Mohamed Benbouzid
University of Brest
France

Yassine Amirat
ISEN Yncréa Ouest
France

Elhoussin Elbouchikhi
AISEN Yncréa Ouest
France

Editorial Office
MDPI
St. Alban-Anlage 66
4052 Basel, Switzerland

This is a reprint of articles from the Special Issue published online in the open access journal *Energies* (ISSN 1996-1073) (available at: https://www.mdpi.com/journal/energies/special_issues/Marine_Energy_Converters).

For citation purposes, cite each article independently as indicated on the article page online and as indicated below:

LastName, A.A.; LastName, B.B.; LastName, C.C. Article Title. *Journal Name* **Year**, *Article Number*, Page Range.

ISBN 978-3-03928-278-4 (Pbk)
ISBN 978-3-03928-279-1 (PDF)

© 2020 by the authors. Articles in this book are Open Access and distributed under the Creative Commons Attribution (CC BY) license, which allows users to download, copy and build upon published articles, as long as the author and publisher are properly credited, which ensures maximum dissemination and a wider impact of our publications.
The book as a whole is distributed by MDPI under the terms and conditions of the Creative Commons license CC BY-NC-ND.

Contents

About the Special Issue Editors vii

Preface to "Marine Tidal and Wave Energy Converters: Technologies, Conversions, Grid Interface, Fault Detection, and Fault-Tolerant Control" ix

Khalil Touimi, Mohamed Benbouzid and Zhe Chen
Optimal Design of a Multibrid Permanent Magnet Generator for a Tidal Stream Turbine
Reprinted from: *Energies* **2020**, *13*, 487, doi:10.3390/en13020487 1

Milu Zhang, Tianzhen Wang, Tianhao Tang, Zhuo Liu and Christophe Claramunt
A Synchronous Sampling Based Harmonic Analysis Strategy for Marine Current Turbine Monitoring System under Strong Interference Conditions
Reprinted from: *Energies* **2019**, *12*, 2117, doi:10.3390/en12112117 21

Stephanie Ordonez-Sanchez, Matthew Allmark, Kate Porter, Robert Ellis, Catherine Lloyd, Ivan Santic, Tim O'Doherty, Cameron Johnstone
Analysis of a Horizontal-Axis Tidal Turbine Performance in the Presence of Regular and Irregular Waves Using Two Control Strategies
Reprinted from: *Energies* **2019**, *12*, 367, doi:10.3390/en12030367 35

James Kelly, Endika Aldaiturriaga and Pablo Ruiz-Minguela
Applying International Power Quality Standards for Current Harmonic Distortion to Wave Energy Converters and Verified Device Emulators
Reprinted from: *Energies* **2019**, *12*, 3654, doi:10.3390/en12193654 57

Marios Charilaos Sousounis and Jonathan Shek
Wave-to-Wire Power Maximization Control for All-Electric Wave Energy Converters with Non-Ideal Power Take-Off
Reprinted from: *Energies* **2019**, *12*, 2948, doi:10.3390/en12152948 79

Brenda Rojas-Delgado, Monica Alonso, Hortensia Amaris and Juan de Santiago
Wave Power Output Smoothing through the Use of a High-Speed Kinetic Buffer
Reprinted from: *Energies* **2019**, *12*, 2196, doi:10.3390/en12112196 107

Mohd Nasir Ayob, Valeria Castellucci, Johan Abrahamsson and Rafael Waters
A Remotely Controlled Sea Level Compensation System for Wave Energy Converters
Reprinted from: *Energies* **2019**, *12*, 1946, doi:10.3390/en12101946 135

Xu Wang and Yanxia Shen
Fault Tolerant Control of DFIG-Based Wind Energy Conversion System Using Augmented Observer
Reprinted from: *Energies* **2019**, *12*, 580, doi:10.3390/en12040580 151

About the Special Issue Editors

Mohamed Benbouzid received a B.Sc. degree in electrical engineering from the University of Batna, Batna, Algeria, in 1990, M.Sc. and Ph.D. degrees in electrical and computer engineering from the National Polytechnic Institute of Grenoble, Grenoble, France, in 1991 and 1994, respectively, and the Habilitation à Diriger des Recherches degree from the University of Picardie "Jules Verne," Amiens, France, in 2000. After receiving his Ph.D. degree, he joined the Professional Institute of Amiens, University of Picardie "Jules Verne," where he was an Associate Professor of Electrical and Computer Engineering. Since September 2004, he has been with the University of Brest, Brest, France, where he is a Full Professor of Electrical Engineering. Prof. Benbouzid is also a Distinguished Professor and a 1000 Talent Expert at the Shanghai Maritime University, Shanghai, China. His main research interests and experience include analysis, design, and control of electric machines, variable-speed drives for traction, propulsion, and renewable energy applications, and the fault diagnosis of electric machines. Prof. Benbouzid has been elevated as an IEEE Fellow for his contributions to diagnosis and fault-tolerant control of electric machines and drives. He is also a Fellow of the IET. He is the Editor-in-Chief of the International Journal on Energy Conversion and the Applied Sciences (MDPI) Section on Electrical, Electronics and Communications Engineering. He is a Subject Editor for *IET Renewable Power Generation*. He is also an Associate Editor of *IEEE Transactions on Energy Conversion*.

Yassine Amirat received B.Sc. and M.Sc. degrees in electrical engineering from the University of Annaba, Annaba, in 1994 and 1997, respectively. He was a lecturer at Annaba University from 2000 to 2010. He obtained a Ph.D. degree in wind turbine condition monitoring at the University of Brest, Brest, France in 2011. He is currently an Associate Professor of Electrical Engineering at ISEN Yncréa Ouest, Brest, France. He is also an affiliated member of the Institut de Recherche Dupuy de Lôme (UMR CNRS 6027). His main research interests include electrical machine fault detection and diagnosis, fault-tolerant control, and signal processing and statistics for power systems monitoring. He is also interested in renewable energy applications: wind turbines, marine current turbines, and hybrid generation systems. Dr. Amirat is an IEEE Senior Member. He is an Associate Editor of Springer's journal *Electrical Engineering* and MDPI's *Journal of Marine Science and Engineering*.

Elhoussin Elbouchikhi received a diploma engineer degree (Dipl.-Ing.) in automatic and electrical engineering and a research Master's degree in automatic systems, computer science and decision from the National Polytechnic Institute of Toulouse (INPENSEEIHT), Toulouse, France, in 2010, and a Ph.D degree in 2013 from the University of Brest, Brest, France. After receiving his Ph.D. degree, he was a Post-Doctoral Researcher at ISEN Yncréa Ouest, Brest, France and an Associate Member of the LBMS Laboratory (EA 4325) from October 2013 to September 2014. Since September 2014, he has been an Associate Professor at ISEN Yncréa Ouest, Brest, France and is an affiliated member of the Institut de Recherche Dupuy de Lôme (UMR CNRS 6027). His main current research interests include electrical machine fault detection and diagnosis, fault-tolerant control in marine current turbines, and signal processing and statistics for power systems monitoring. He is also interested in energy management systems in microgrids and renewable energy applications such as marine current turbines, wind turbines, and hybrid generation systems. Dr. Elbouchikhi is an IEEE Senior Member. He is a Topic Editor for the MDPI journal *Energies*.

Preface to "Marine Tidal and Wave Energy Converters: Technologies, Conversions, Grid Interface, Fault Detection, and Fault-Tolerant Control"

The worldwide potential of electric power generation from marine tidal currents, waves, or offshore winds is enormous. The high load factor resulting from the fluid properties and the predictable resource characteristics make tidal and wave energy resources attractive and advantageous for power generation and advantageous when compared to other renewable energies. The technologies are just beginning to reach technical and economic viability to make them potential commercial power sources in the near future. While only a few small projects currently exist, the technology is advancing rapidly and has huge potential for generating bulk power. Moreover, international treaties related to climate control and dwindling fossil fuel resources have encouraged us to harness energy sustainably from such marine renewable sources. Several demonstrative projects have been scheduled to capture tidal and wave energies. A number of these projects have now reached a relatively mature stage and are close to completion. However, very little is known to the academic world about these technologies beyond the basics of their energy conversion principles. While research emphasis is more towards hydrodynamics and turbine design, very limited activities are witnessed in power conversion interface, control, and power quality aspects. Regarding this emerging and promising area of research, this book aims to present recent results, serving to promote successful marine renewable energies integration to the grid or to standalone microgrids.

Mohamed Benbouzid, Yassine Amirat, Elhoussin Elbouchikhi
Special Issue Editors

Article

Optimal Design of a Multibrid Permanent Magnet Generator for a Tidal Stream Turbine

Khalil Touimi [1,2], Mohamed Benbouzid [1,3],* and Zhe Chen [4]

1. Institut de recherche Dupuy de Lôme (UMR CNRS 6017 IRDL), University of Brest, 29238 Brest, France; Khalil.Touimi@univ-brest.fr
2. École Militaire Polytechnique, 16111 Alger, Algeria
3. Logistics Engineering College, Shanghai Maritime University, Shanghai 201306, China
4. Department of Energy Technology, Aalborg University, 9220 Aalborg, Denmark; zch@et.aau.dk
* Correspondence: Mohamed.Benbouzid@univ-brest.fr; Tel.: +33-2980-18007

Received: 15 November 2019; Accepted: 16 January 2020; Published: 19 January 2020

Abstract: Tidal stream energy is acquiring more attention as a future potential renewable energy source. Considering the harsh submarine environment, the main challenges that face the tidal stream turbine (TST) industry are cost and reliability. Hence, simple and reliable technologies, especially considering the drivetrain, are preferred. The multibrid drivetrain configuration with only a single stage gearbox is one of the promising concepts for TST systems. In this context, this paper proposes the design optimization of a multibrid permanent magnet generator (PMG), the design of a planetary gearbox, and afterwards analyzes the multibrid concept cost-effectiveness for TST applications. Firstly, the system analytical model, which consists of a single-stage gearbox and a medium speed PMG, is presented. The optimization methodology is afterwards highlighted. Lastly, the multibrid system optimization results for different gear ratios including the direct-drive topology are discussed and compared where the suitable gear ratio (topology) is investigated. The achieved results show that the multibrid concept in TST applications seems more attractive than the direct-drive one especially for high power ratings.

Keywords: tidal stream turbine; multibrid concept; direct-drive; permanent magnet generator; single stage gearbox; design optimization

---

1. Introduction

Tidal stream energy is one of the promising renewable energy sources, which is highly predictable and its potential can exceed 120 GW [1,2]. It is mainly harnessed by horizontal axis turbines where the marine current kinetic power is converted into an electrical one. Despite the infancy of tidal stream turbine (TST) technologies, various machines and prototypes have been developed in recent decades, and different concepts are competing for supremacy [3–5]. In addition to the technology infancy, the harsh submarine environment increases the criticality of the TST subsystems. Therefore, the main challenges that face the tidal stream turbine industry are the energy cost and the system reliability, which means simple and reliable technologies should be adopted. Drivetrain and generator topology choice typically affects the availability of the system as well as the produced energy cost. The main TST configuration types are gearless TST (direct-drive), mechanically geared TST, and magnetically geared TST [6–8].

Direct-drive topology (Figure 1), which was designed to avoid gearbox failures in wind turbines, is attractive due to its simplicity and its high reliability. However, direct-drive generators are non-standard electric machines and have some disadvantages such as the heavy weight, large diameter, and therefore high cost. Geared generators are compact, robust, and economically available compared to direct-drive ones, in addition to the fact that mechanical gearbox technologies are mature. Moreover, authors in [9] addressed the criticality of wind turbine subsystems in different sites and provided a comparison between geared and direct-drive wind turbines. The study shows that direct-drive systems are less reliable than geared ones. However, mechanical gearboxes are still critical subsystems, which can lower tidal stream turbines availability [10]. In [11] a comparative study between direct-drive tidal stream turbines and gearbox driven ones has been carried out, where the result suggested the multibrid concept as an alternative compact drivetrain for TST applications in terms of reliability and availability.

Figure 1. OpenHydro/Naval Energies direct-drive tidal stream turbine [4].

The multibrid configuration (Figure 2a), with a single-stage planetary gearbox associated to a medium speed PMG, combines the advantages of both geared and gearless drivetrain [12]. Indeed, a medium speed generator is cheaper and more efficient than a direct-drive one and a single-stage gearbox is lighter and more reliable than a multiple-stage one. Wind turbines manufacturers have developed multibrid technology such as Multibrid (M5000) and WinWind (WWD-3) [13] (Figure 2b), [14]. The same concept has been designed, realized, and tested for a small scale TST system in the Chinese Zhoushan water channel [15] (Figure 3).

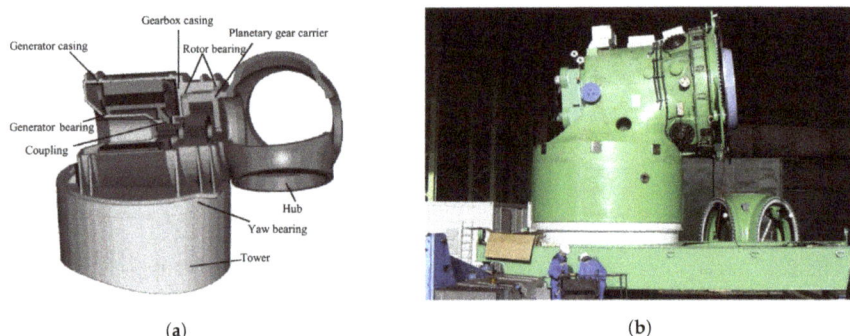

(a) (b)

Figure 2. The multibrid concept: (a) Schematic illustration [12], (b) The AREVA Multibrid M5000 5 MW wind turbine nacelle [11].

The cost of a Multibrid TST depends on the gearbox ratio and the generator diameter. Gearboxes with high gear ratio are heavier and more expensive. However, their high-speed output leads to cheaper generators with low diameter. Hence, to determine the appropriate gearbox ratio and generator dimensions for given specifications, a system optimization is required by minimizing its active parts cost.

Figure 3. Small scale multibrid tidal stream turbine [15].

Previous studies on wind turbine systems compared geared generators (including single stage gearbox driven ones) to direct-drive generators in terms of cost. In [16], the authors compared quantitatively different drive-train configurations and different generator topologies. In this study, which highlights the multibrid concept, the design of the generators is not optimized and the gear ratio is chosen in advance. In [17] and based on [16], the authors have estimated and compared the cost of energy of different drive-train configurations. Unlike the above-cited papers, Hui et al. [12] have investigated gearbox ratios and power ratings cost-effective ranges of multibrid permanent magnet wind generators including direct-drive ones. Concerning TST design optimization, in [5,18], the authors compared different optimized direct-drive PMG topologies (rim-driven vs. pod assembly and radial flux vs. axial flux PMG). However, for the geared drive-train configuration especially, the multibrid concept was not considered. Even if the wind turbine systems seem similar to TST ones, some fundamental differences on design and operation require more investigation, such as biofouling and marine current turbulence [19,20]. Therefore, both the blades and yaw pitch subsystems are avoided due to their high criticality in such a hostile environment [9].

In this paper, a design optimization of PMG for TST system is proposed in order to analyze the cost-effectiveness of the multibrid concept and compare it to the direct-drive one. In this context, the Multibrid TST analytical model is presented and it consists of: the turbine model, the single stage planetary gearbox model, the three-phase PMG two-dimensional (2D) electromagnetic model, and the power electronics converter model. Figure 4 is therefore illustrating a grid-connected single stage gearbox driven PMG, highlighting each subsystem. The proposed design optimization process is performed using the interior-point method to minimize the active material cost of the generator. The suitable drivetrain configuration is afterwards investigated for different power ratings (500 kW, 1.5 MW, and 5 MW) and the achieved results are compared and discussed.

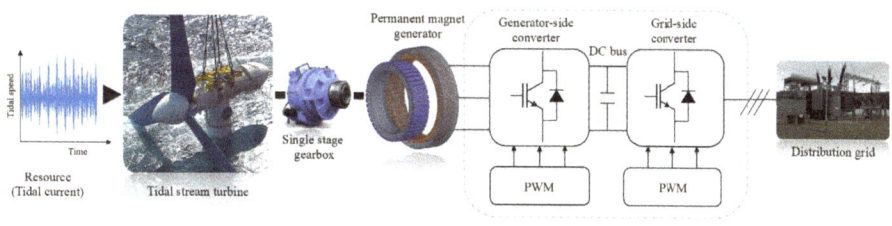

Figure 4. Scheme of a grid-connected single stage permanent magnet generator-based tidal stream turbine.

2. System Modeling

2.1. Renewable Resource and Tidal Turbine Modeling

Tidal current velocity data in the site near Ouessant Island were collected by the French navy hydrographic and oceanographic service [21]. The amplitude and direction measurements of the tidal current velocity are done hourly during one year (8760 h). The optimal direction, which provides the maximum of energy, is calculated as described in [22] (Figure 5). Tidal speed through the optimal direction is shown in Figure 6.

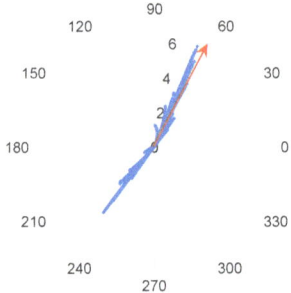

Figure 5. Tidal velocity in polar coordinates.

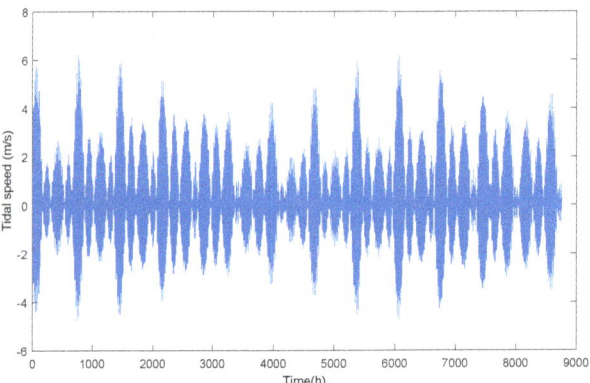

Figure 6. Tidal velocity in the Ouessant Island.

2.1.1. Power and Energy Calculation

Energy calculation is done considering the optimal angle direction (61°). Concerning the input shaft power from a tidal turbine, it is calculated as a function of tidal currents speed and the turbine rotor diameter.

$$P_T = \frac{1}{2}\rho C_p(\lambda, \beta) A_t v^3, \tag{1}$$

where $A_t = \frac{1}{4}\pi D^2$ is the turbine blade swept area, ρ is the sea water density, $C_p(\lambda, \beta)$ is the power coefficient which is a function of tip speed ratio (λ) and the pitch angle of the tidal turbine blades (β).

The annual energy production (AEP) can be calculated by summing the harnessed energy in each hour (the tidal current speed is assumed non-variable). Figure 7 shows the energy distribution according to tidal current speed amplitude in the Ouessant site.

$$AEP = \int_{v_i}^{v_n} P_T(|v|)OCC(|v|)dv + P_{Tr}\int_{v_n}^{v_c} OCC(|v|)dv, \tag{2}$$

where v_i is the cut-in tidal current speed, v_c is the cut-out tidal current speed, v_n is the rated tidal current speed, and P_{Tr} is the rated input shaft power.

The $OCC(|v|)$ function represents the tidal current speed amplitude distribution (Figure 8).

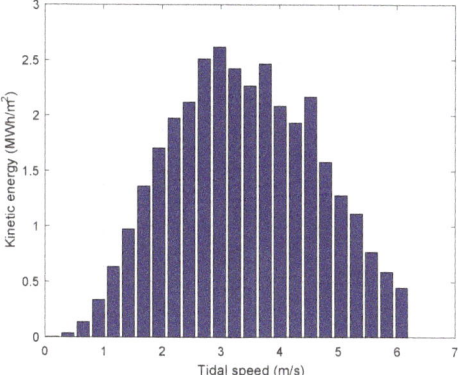

Figure 7. Tidal current energy distribution.

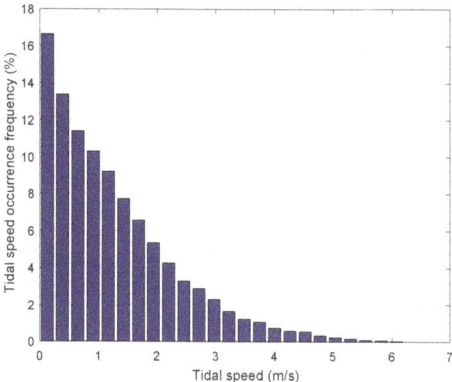

Figure 8. Tidal current amplitude speed distribution.

2.1.2. Power Rating Choice

To harness all the energy, a maximal power should be chosen as a rated one and an oversized system will be required. In wind turbines, mechanical limitation of power using blade or yaw pitch subsystems are adopted. However, due to the harsh submarine conditions such mechanical subsystems should be avoided. An alternative using an over-speed power limitation is adopted in this study. In this context, when the input power is less than the rated one, the power coefficient is maintained at his maximum ($C_{pmax} = 0.455$) (Table 1). However, when the input power is higher than the rated one, the generator accelerates and reduces the power coefficient. This methodology is detailed in [8]. The power rating (limitation power) is chosen around 30% of the maximum power where 90% of the total energy can be harnessed (Figure 9). As the swept area considered is 1 m^2, the same power ratio of 30% is maintained, and for each power rating (500 kW, 1.5 MW, 5 MW) only the blade's diameter (swept area) is calculated.

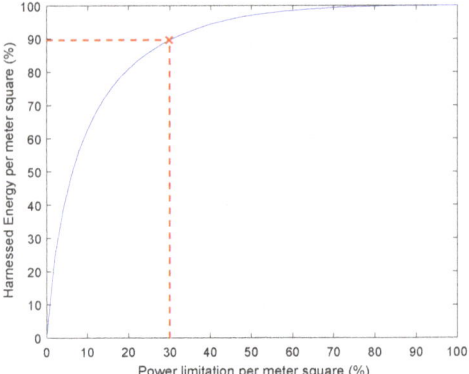

Figure 9. Harnessed energy rate versus power limitation rate.

2.2. Gearbox Modeling

The gearbox converts the turbine rotor slow rotational speed and high shaft torque to high rotational speed and low torque. The more its gear ratio is high, the more its cost and weight increases. However, the opposite happens to the generator because its input shaft torque decreases. It exists two main types of gear trains: parallel shaft and planetary. In this study, a planetary single stage gearbox is considered due to its high power density (Figure 10). The volume of each part of the planetary gearbox is estimated to calculate its total mass [23,24].

$$\sum FWd^2 = FWd_s^2 + FWd_p^2 + K_r FWd_r^2, \tag{3}$$

where FW is the face width of the gear, d_s, d_p, and d_r are the diameter of the sun, the planets, and the ring gear respectively. FWd^2 presents the gear volume and $K_r = 0.4$ is a scaling factor and it is selected from [23,24].

The weight of the planetary single stage gearbox is a function of the gear ratio and the transmitted shaft torque [25]. Equation (4) is developed to obtain the planetary gearbox total weight.

$$G_{gear} = \frac{W_c}{36050} \frac{2(10^3)T_m K_{ag}}{K_f} \left[\frac{1}{Z} + \frac{1}{Zr_{sn}} + r_{sn} + r_{sn}^2 + K_r \frac{(1+(r_{sn})^{-1})}{Z}(r_{ratio} - 1)^2 \right] \tag{4}$$

In the last Equation (4), T_m is the gearbox output shaft torque and K_{ag} is the application factor. It is chosen from [23] among different application factors. In fact, a factor of 1.0 is chosen when we have a perfectly smooth turbine driving a perfectly smooth generator always at a constant speed (no frictions and no vibrations). Because of the high torque fluctuations due to the high marine energy density [26], an application factor of 1.5 is chosen. K_f is an index of tooth loads intensity and it is empirically estimated from [23]. W_c, which is the weight constant, is also estimated from [23]. r_{ratio} is the single stage gearbox ratio (between the carrier and the sun gear), $r_{sn} = (r_{ratio}/2) - 1$ is the ratio between the sun and planet gears, and Z is the number of planet gears (Table 1).

The estimated cost of the single stage gearbox is given as

$$C_{gear} = c_{gear} G_{gear}, \tag{5}$$

where c_{gear} is the specific cost of the single stage gearbox (Table 1).

Concerning losses, only speed dependent ones are considered (seal losses and lubricant losses). Regarding power-dependent losses, they are negligible compared to speed dependent ones [27].

$$p_{gear} = k_g P_N \frac{n_r}{n_{r_N}} \quad (6)$$

In the precedent Equation (6), k_g is the speed-dependent losses constant, P_N is the rated power of the TST, n_r is the rotor speed, and n_{r_N} is the rated rotor speed.

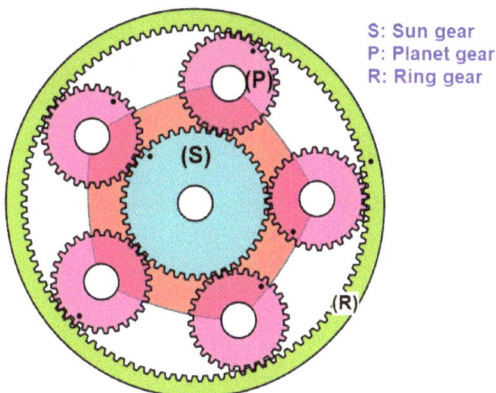

Figure 10. Illustration of a planetary gearbox with five planet gears.

2.3. Single Stage Geared PMG Design

The generator considered in this paper is a three phase radial flux permanent magnet one [8]. Figure 11 shows the geometric parameters of one pair of poles and its structure. The magnets are surface mounted and the generator curvature is assumed insignificant. For design purposes, the adopted modeling is a 2D analytical electromagnetic model based on magnetic circuit calculation [18,28]. The objective is to calculate the size of the generator knowing its specifications.

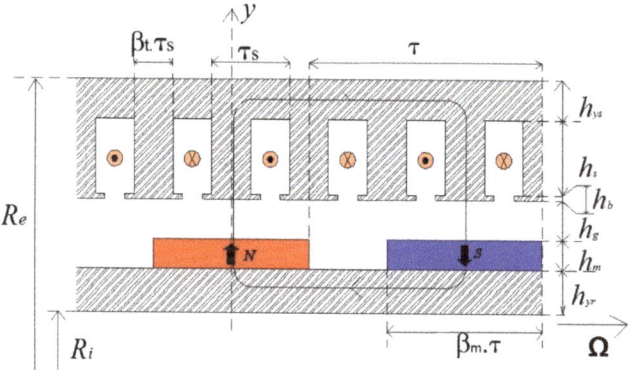

Figure 11. Basic dimensions of one pair of poles [8].

2.3.1. Electromagnetic Torque

The average electromagnetic torque results from the interaction between the fundamental electromotive forces and the currents of the phases (considered sinusoidal) at the nominal operating point [29].

$$<T_{EM}> = 4\sqrt{2} A_L k_{b1} B_{g_{max}} R_s^2 L_m \xi_{3D} \sin(\beta_m \frac{\pi}{2}) |\cos(\psi)|, \quad (7)$$

where A_L is the current loading in the stator, B_{gmax} is the maximum air-gap flux density under the magnet, k_{b1} is the first harmonic winding factor, ψ is the phase shift between the fundamental of the electromotive force and the current, R_s is the stator radius, L_m is the equivalent core length, and ξ_{3D} is a corrective coefficient which takes into consideration the 3D flow leakage.

2.3.2. Air-Gap

The mechanical air-gap is given by the following empirical formula [30]

$$h_g = 2k_D R_s, \tag{8}$$

where k_D is a coefficient, which considers the deformations caused by the forces acting on the rotating rotor.

The additional Carter air-gap $h_{g'}$ is calculated as [30]

$$h_{g'} = (k_c - 1)\left(h_g + \frac{h_m}{\mu_{rm}}\right), \tag{9}$$

where k_c is the Carter factor, h_m is the magnet height, and μ_{rm} is the magnets relative permeability.

2.3.3. Magnet Height

The magnet height model calculation considers inter-polar 2D leakage flow [28]

$$h_m = \frac{\tau}{2\pi}\left[\ln\left(\frac{(\mu_{rm}+1)B_{gmax}\exp\frac{-\pi}{\tau}(h_g+h_{g'}) - \frac{(\mu_{rm}-1)}{(\mu_{rm}+1)}\exp\frac{\pi}{\tau}(h_g+h_{g'}) - 2B_r}{(\mu_{rm}+1)B_{gmax}\exp\frac{\pi}{\tau}(h_g+h_{g'}) - \frac{(\mu_{rm}-1)}{(\mu_{rm}+1)}\exp\frac{-\pi}{\tau}(h_g+h_{g'}) - 2B_r}\right)\right], \tag{10}$$

where B_r is the magnet's remanent flux density and B_{gmax} is the maximum air-gap flow density, and τ is the pole pitch.

2.3.4. Slot Height

The slot height depends on the current loading A_L, the fill factor k_f, and the teeth pitch ratio β_t.

$$h_s = \frac{A_L}{k_f J(1-\beta_t)} \tag{11}$$

2.3.5. Stator and Rotor Yoke Height

The stator yoke height h_{ys} is determined in a way to avoid its saturation. With the same principle, the rotor yoke height h_{yr} is developed [28].

$$h_{ys} = \beta_m \frac{\pi R_s}{2p} \frac{B_{gmax}}{B_{sat}} + \frac{1}{3} \frac{\mu_0 \mu_{rm} \sqrt{2} A_L \pi^2 R_s^2}{(h_m + \mu_{rm}(h_g + h_{g'}))S_{pp}mp^2 B_{sat}}, \tag{12}$$

$$h_{yr} \approx h_{ys} \tag{13}$$

where S_{pp} is the number of slots per pole per phase, m is the phases number, and p is pole pairs number.

2.3.6. Teeth Pitch Ratio

The teeth pitch ratio β_t is calculated in a way to assure a non saturation of the generator when it is over-fluxed ($\psi = \pi/2$) and the air-gap flow density is at its maximum $Bg = B_{gmax}$ along the pole [28].

$$\beta_t = \frac{B_{gmax}}{B_{sat}} + \frac{\mu_0 \mu_{rm} \sqrt{2} A_L \pi R_s}{(h_m + \mu_{rm}(h_g + h_{g'}))S_{pp}mp B_{sat}} \tag{14}$$

2.3.7. Maximum Magnetic Field

To avoid the irreversible permanent magnet demagnetization, the maximum magnetic field H_{max} has to be less than the PM coercive magnetic field H_{cj}. The maximum magnetic field is calculated in the worst scenario where stator flow density is opposite to the rotor flow density. H_{max} will be introduced as a constraint in the optimization process [28].

$$|H_{max}| = \frac{\sqrt{2}A_L \pi R_s}{(h_m + \mu_{rm}(h_g + h_{g'}))S_{pp}mp} + \frac{(h_g + h_{g'})B_{gmax}}{\mu_0 h_m} \tag{15}$$

2.3.8. Iron Losses

The specific iron losses are estimated by using the Steinmetz formula [31,32]

$$p_{Fe} = 2p_{Fe0h}(\frac{f_e}{f_0})(\frac{\widehat{B}_{Fe}}{\widehat{B}_0})^2 + 2p_{Fe0e}(\frac{f_e}{f_0})^2(\frac{\widehat{B}_{Fe}}{\widehat{B}_0})^2, \tag{16}$$

where f_e is the field frequency in the iron, p_{Fe0h} represents the specific hysteresis loss, p_{Fe0e} represents the specific eddy current loss in the laminated stator core for a frequency f_0 of 50 Hz and a flux density $\widehat{B}_0$ of 1.5 T.

2.3.9. Synchronous Inductance

The synchronous inductance L_s is the sum of the magnetizing inductance L_{sm} and the leakage inductance L_{sl}. It is calculated as [30]

$$L_{sm} = \frac{6\mu_0 L_e R_s (k_w N_s)^2}{\pi p^2 (h_g + h_{g'})}, \tag{17}$$

where μ_0 is the vacuum permeability constant and N_s is the number of turns of the phase winding.

Only slot leakage and the end-winding leakage inductances are considered to calculate the leakage inductance L_{sl}. Skew leakage inductance is ignored because the stator slots are not skewed.

2.4. Power Electronic Converter Design

A two level back-to-back pulse width modulation (PWM) full scale converter is used to inject power from the generator to the grid. Its specific cost estimate is presented in Table 1. Concerning losses rate, they are considered to be about 3% at the rated load [33].

Table 1. Modeling parameters of the tidal stream turbine system.

Tidal Stream Turbine			
Rated power P_N [MW]	0.5	1.5	5
Rated rotor speed n_{r_N} [rpm]	80.3	47.0	25.8
Rotor diameter D [m]	6	10.3	18.8
Cut it tidal current speed v_i [m/s]		1.0	
Cut out tidal current speed v_{out} [m/s]		6.2	
Maximum power coefficient C_{pmax}		0.455	
Optimum tip speed ratio λ_{opt}		5.90	
Sea water density [kg/m³]		995.6	

Table 1. *Cont.*

Single Stage Planetary Gearbox	
Gearbox application factor K_{ag}	1.5
K-factor K_f [N/mm^2]	2.76
Gearbox weight constant W_c	0.6
Planet gears number Z	6
Gearbox specific cost c_{gear} [€/kg]	6
Speed dependent losses constant k_g [%]	1.5
PMG System	
Hysteresis losses at 1.5 T and 50 Hz p_{Fe0h} [W/kg]	2
Eddy-current losses at 1.5 T and 50 Hz p_{Fe0e} [W/kg]	0.5
Specific cost of electrical steel c_{Fe} [€/mT]	449.77
Specific cost of copper c_{Cu} [€/mT]	4259.18
Specific cost of NdFeB magnet c_m [€/mT]	84,538.60
Specific cost of power electronics c_{conv} [€/kW]	40

3. Design Optimization

The design optimization is based on the modeling of each part of the system (the turbine, the gearbox, the power electronics converters, and the PMG generator). The electromagnetic specifications are all the fixed design parameters, the gearbox ratio is chosen before the optimization process starting, and tidal power specifications are calculated from tidal current speed data. The analytical model can be represented by a non-linear function, which has as input a geometrical design variables vector. To evaluate the cost-function, an iterative inversion of the non-linear model is adopted. By considering the constraint, the interior-point optimization technique is used to minimize the cost-function. The optimal design is the one which has the lowest cost (Figure 12).

3.1. Cost-Function

The TST cost depends on its active materials weight, its structure cost, and its manufacturing cost. In this study, only the generator's active materials cost is considered in addition to the gearbox and the power electronic converter costs. The active material cost depends on the size of each material. Hence, the generator cost is calculated only from the size parameters $G = (L_m, hs, hm, hys, \beta_t)$.

$$C_g = c_{Cu}G_{Cu} + c_{Fe}G_{Fe} + c_m G_m \tag{18}$$

$$C_{TST} = C_g + C_{gear} + C_{conv}, \tag{19}$$

where C_{gear} is the single stage gearbox cost, C_{conv} is the estimated power electronics cost, c_{Cu}, c_{Fe}, c_m are the copper, the iron, and the permanent magnet specific costs, and G_{Cu}, G_{Fe}, G_m are the copper, the iron, and the permanent magnet weights, respectively.

The cost-function depends on the gearbox ratio and the size parameters of the PMG. However, if the analytical modeling is considered, the PMG cost will depend only on the following variables $X = (A_L, J, B_{gmax}, p, R_s)$.

$$X^* = \min_{X \in \mathbb{D}} ||C_g(X)||, \tag{20}$$

where X is the vector including the independent variables $(A_L, J, B_{gmax}, p, R_s)$ and $\mathbb{D}$ is the set of possible solutions.

3.2. Optimization Constraints

The optimization is performed under electromagnetic and mechanical constraints, which define the $\mathbb{D}$ set.

The first constraint on the pole pair number is related to the maximum electrical frequency. In fact, a high pole pair number leads to a higher electrical frequency resulting in high iron losses. To avoid this, the maximum electrical frequency (f_{max}) allowed in laminated steel core is limited, which can be considered as a limitation of the pole pair number. The second constraint is related to the ratio of slot depth to slot width. This ratio must be in the range of 4–10 to avoid excessive mechanical vibrations [12,34]. This limitation is converted to two additional constraints on the pole pair number.

The maximum magnetic field $H_{max}(X)$ (see Equation (15)) is limited to be smaller than the permanent magnet coercive field to prevent demagnetization. The current density J and the loading current A are limited in the range of 3–6 A/mm^2 and 40–60 kA/m respectively [12]. The generator efficiency is considered to be greater than 0.96. On the other hand, the phase voltage root mean square is fixed to 690 V.

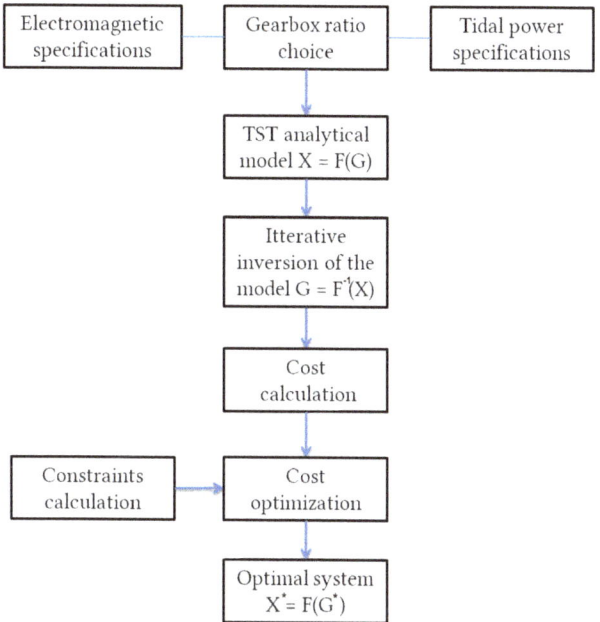

Figure 12. Flowchart describing the design optimization procedure.

4. Design Results and Discussion

To investigate the cost-effectiveness of the multibrid concept in TST systems, design optimization of the system is performed for different gearbox ratios (1:1,3:1,5:1,7:1,9:1,11:1) and different power ratings (0.5 MW, 1.5 MW, 5 MW). The total harnessed energy in the Ouessant site is calculated for each power rating. According to the optimization results, the gear ratio (3:1) is the optimal one whatever the power level is (Figure 13). When it is less than (5:1), geared systems are cheaper than direct-drive ones. Concerning the estimated TST cost per kWh (Figure 14), the 1.5 MW and 5 MW TST energy is cheaper than the 500 kW TST one. It seems that a rated power around 1.5 MW for the Ouessant site is more preferable than very high power ratings.

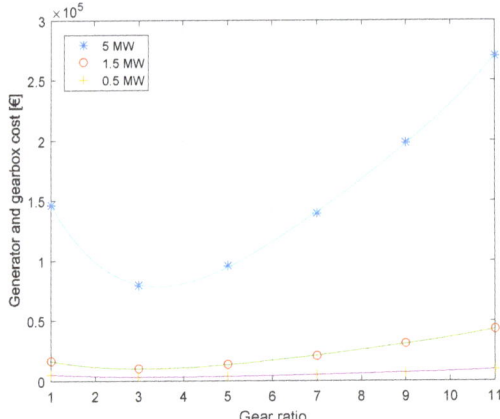

Figure 13. Tidal stream turbine (TST) estimated cost for different gear ratios and different power ratings.

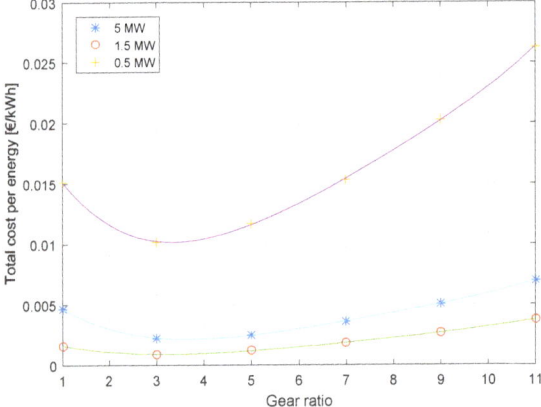

Figure 14. TST total estimated cost per kWh.

As shown in Figure 15, the generator and gearbox cost estimations are separately presented at the power of 1.5 MW. For high gear ratios the gearbox cost increases exponentially while the generator cost decreases slightly. The generator cost is however highly reduced when the gearbox ratio changes from (1:1) (direct-drive) to (3:1). The comparison shows that gearbox low ratios are more interesting especially when the power rating is high but for higher gear ratios the direct-drive configuration is preferable.

The second part considers the 1.5 MW generator's active materials cost and weight. Figure 16 shows that the direct-drive system is the heaviest one compared to the other configurations. On the other hand, the weight of the (3:1) geared generator is around 35% of the total weight of the direct-drive generator.

Regarding PMG active materials cost (Figure 17), the direct-drive configuration is the most expensive and permanent magnet's cost is extremely high compared to the other material's costs. However, the multibrid generator's active materials costs are more balanced especially when the gear ratio is high. In Figure 18 the direct-drive configuration cost is taken as a reference to be compared to the other configurations cost. As it is shown, the generator's active materials cost is reduced by around 65% when adopting a (3:1) gearbox. For higher gear ratios, cost reduction can reach approximately 85%.

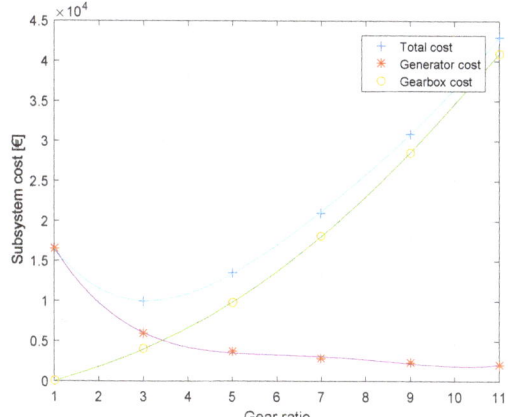

Figure 15. Gearbox and generator costs for different gear ratios at the power rating of 1.5 MW.

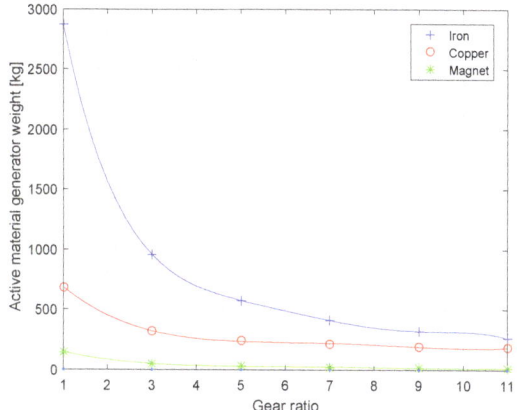

Figure 16. Generator active materials weight for different gear ratios at the power rating of 1.5 MW.

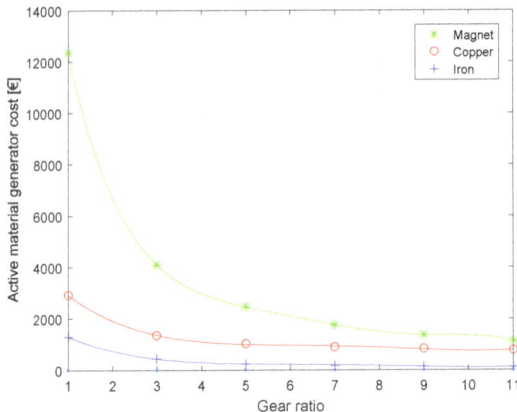

Figure 17. Generator active materials cost for different gear ratios at the power rating of 1.5 MW.

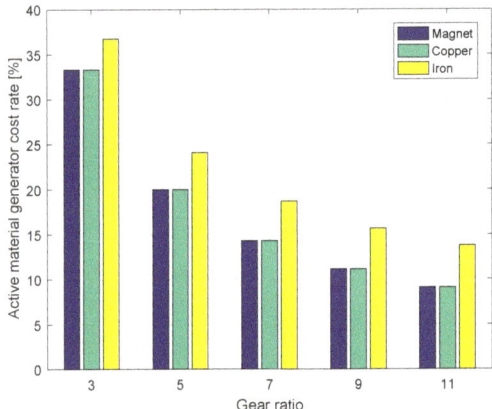

Figure 18. Active material geared generator cost compared to direct drive (DD) generator.

Figure 19 presents four poles of the designed (3:1) geared generator and Figure 20 shows a front and lateral view of the same designed generator at the rating power of 1.5 MW. The two figures give a vision of the designed generator structure and size.

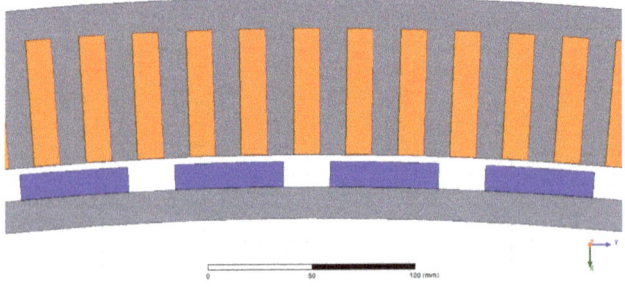

Figure 19. View of the designed (3:1) geared generator at the power rating of 1.5 MW.

Figure 20. Front and lateral view of the designed (3:1) geared generator at the power rating of 1.5 MW.

From the above-presented results, it is shown that the Multibrid TSTs can be a promising alternative to direct-drive TSTs in terms of weight and cost. According to the Ouessant site energy potential, TSTs power ratings of around 1.5 MW are interesting especially if we consider the limited deepness of the site that limits the turbine diameter. A power rating of 5 MW requires a turbine diameter of 18.8 m, which is interesting if the site allows such a size.

For more accurate estimated cost, the generator structure and manufacturing costs should be considered. Moreover, the TST foundations are not addressed in this study and the gearbox is not deeply designed, which could affect the optimization results. However, these comparative study results could be useful for TST designers and could give them insight on the relevance of the multibrid concept.

5. Conclusions

This paper investigated the application of the multibrid concept for tidal stream turbines. In this context, a design optimization of multibrid permanent magnet generator has been proposed and the system cost-effectiveness has been analyzed by considering the Ouessant site potential energy. A planetary gearbox rough design is proposed for TST systems and the gearbox weight and cost estimations are presented. Furthermore, the study considers a 2D analytical electromagnetic modeling to size the permanent magnet generator. The achieved optimization results clearly showed that multibrid tidal stream turbines are the solution of choice, in terms of weight and cost, when compared to the direct-drive topology. Otherwise, among multibrid systems, lower gear ratios seem preferable, especially for high power ratings. The cost of the 1.5 MW TST is the lowest regarding its harnessed energy in the Ouessent site and it is not far from the 5 MW one. However, a 1.5 MW TST is preferable if the environment constraints are considered where the site deepness is limited.

Author Contributions: Conceptualization, K.T. and M.B.; Methodology, K.T. and M.B.; Software, K.T.; Validation, K.T., M.B., and Z.C.; Formal Analysis, K.T. and M.B.; Investigation, K.T.; Writing–Original Draft Preparation, K.T.; Writing–Review & Editing, K.T., M.B., and Z.C.; Supervision, M.B. All authors have read and agreed to the published version of the manuscript.

Funding: This research received no external funding

Conflicts of Interest: The authors declare no conflict of interest.

Abbreviations

The following abbreviations are used in this manuscript:

TST	Tidal stream turbine
DD	Direct drive
2D	Two-dimensional
PMG	Permanent magnet generator
3D	Three-dimensional
AEP	Annual energy production
PWM	Pulse width modulation

Nomenclature

P_T	Input shaft power
A_t	Turbine blade swept area
ρ	Sea water density
C_p	Power coefficient
λ	Tip speed ratio
λ_{opt}	Optimum tip speed ratio
β	Pitch angle
v_i	Cut-in tidal current speed
v_c	Cut-out tidal current speed
v_n	Rated tidal current speed
P_{Tr}	Rated input shaft power
OCC	Occurrence frequency
FW	Gear face width
d_s	Sun gear diameter
d_p	Planet gear diameter
d_r	Ring gear diameter
K_r	Scaling factor
T_m	Gearbox output shaft torque
K_{ag}	Application factor
K_f	Tooth loads intensity index
W_c	Gearbox weight constant
r_{ratio}	Gearbox ratio
r_{sn}	Gear ratio between sun and planet gears
Z	Planet gears number
c_{gear}	Gearbox specific cost
G_{gear}	Gearbox weight
C_{gear}	Gearbox estimated cost
p_{gear}	Gearbox losses
k_g	Speed-dependent losses constant
P_N	Tidal stream turbine rated power
n_r	Rotor speed
n_{r_N}	Rated rotor speed
T_{EM}	Electromagnetic torque
A_L	Stator current loading
$B_{g_{max}}$	Maximum air-gap flux density
B_g	Air-gap flux density
B_{max}	Saturation flux density
k_{b1}	First harmonic winding factor
ψ	Phase shift between the electromotive force and the current
R_s	Stator radius
L_m	Equivalent core length

ξ_{3D}	3D flow leakage corrective coefficient
k_D	Air-gap coefficient
h_g	Mechanical air-gap
$h_{g'}$	Additional Carter air-gap
k_c	Carter factor
h_m	Magnet height
μ_0	Vacuum permeability constant
μ_{rm}	Magnets relative permeability
B_r	Magnets remanent flux density
$B_{g_{max}}$	Maximum air-gap flow density
τ	Pole pitch
h_{ys}	Stator yoke height
h_{yr}	Rotor yoke height
h_s	Slot height
k_f	Fill factor
β_t	Teeth pitch ratio
p	Pole pairs number
S_{pp}	Slots per pole per phase number
m	Phases number
H_{max}	Maximum magnetic field in the magnet
H_{cj}	Permanent magnet coercive magnetic field
p_{Fe}	Iron losses
f_e	Magnetic field frequency in the iron
p_{Fe0h}	Specific hysteresis loss
p_{Fe0e}	Specific eddy current loss
N_s	Phase winding number of turns
L_{sl}	Leakage inductance
C_{conv}	Power electronics cost
C_g	Permanent magnet generator cost
C_{TST}	Tidal stream turbine cost
c_{Cu}	Copper specific costs
c_{Fe}	Iron specific costs
c_m	Permanent magnet specific costs
G_{Cu}	Copper specific weight
G_{Fe}	Iron specific weight
G_m	Permanent magnet specific weight
$\mathbb{D}$	Set of possible solutions
f_{max}	Maximum electrical frequency

References

1. Benbouzid, M.; Titah-Benbouzid, H.; Zhou, Z. *Ocean Energy Technologies*; Abraham, M.A., Ed.; Encyclopedia of Sustainable Technologies; Elsevier: Amsterdam, The Netherlands, 2017; pp. 73–85, ISBN 978-0-128-04677-7.
2. Selin, N.E. Tidal Power. April 2019. Available online: https://www.britannica.com/science/tidal-power (accessed on 6 June 2019).
3. Flambard, J.; Amirat, Y.; Feld, G.; Benbouzid, M.; Ruiz, N. River and Estuary Current Power Overview. *J. Mar. Sci. Eng.* **2019**, *7*, 365. [CrossRef]
4. Zhou, Z.; Benbouzid, M.; Charpentier, J.F.; Scuiller, F.; Tang, T. Developments in large marine current turbine technologies–A review. *Renew. Sustain. Energy Rev.* **2017**, *71*, 852–858. [CrossRef]
5. Djebarri, S.; Charpentier, J.F.; Scuiller, F.; Benbouzid, M. Comparison of direct-drive PM generators for tidal turbines. In Proceedings of the 2014 International Power Electronics and Application Conference and Exposition, Shanghai, China, 5–8 November 2014; pp. 474–479.

6. Zeinali, R.; Keysan, O. A Rare-Earth Free Magnetically Geared Generator for Direct-Drive Wind Turbines. *Energies* **2019**, *12*, 447. [CrossRef]
7. Keysan, O.; McDonald, A.S.; Mueller, M. A direct drive permanent magnet generator design for a tidal current turbine (SeaGen). In Proceedings of the 2011 IEEE International Electric Machines & Drives Conference (IEMDC), Niagara Falls, ON, Canada, 15–18 May 2011; pp. 224–229.
8. Djebarri, S.; Charpentier, J.F.; Scuiller, F.; Benbouzid, M. Design methodology of permanent magnet generators for fixed-pitch tidal turbines with overspeed power limitation strategy. *J. Ocean. Eng. Sci.* **2019**. [CrossRef]
9. Ozturk, S.; Fthenakis, V.; Faulstich, S. Failure Modes, Effects and Criticality Analysis for Wind Turbines Considering Climatic Regions and Comparing Geared and Direct Drive Wind Turbines. *Energies* **2018**, *11*, 2317.
10. Touimi, K.; Benbouzid, M.; Tavner, P. A Review-based Comparison of Drivetrain Options for Tidal Turbines. In Proceedings of the 2018 IEEE International Power Electronics and Application Conference and Exposition (PEAC), Shenzhen, China, 4–7 November 2018; pp. 1–6.
11. Touimi, K.; Benbouzid, M.; Tavner, P. Tidal stream turbines: With or without a Gearbox? *Ocean. Eng.* **2018**, *170*, 74–88. [CrossRef]
12. Li, H.; Chen, Z.; Polinder, H. Optimization of multibrid permanent-magnet wind generator systems. *IEEE Trans. Energy Convers.* **2009**, *24*, 82–92. [CrossRef]
13. Detailed Technical-Specification WWD-3. Available online: http://www.ecosource-energy.bg/uploads/Technical_Specification_WWD3.pdf (accessed on 6 June 2019).
14. Multibrid M5000. Available online: https://en.wind-turbine-models.com/turbines/22-multibrid-m5000 (accessed on 6 June 2019).
15. Xu, Q.; Li, W.; Lin, Y.; Liu, H.; Gu, Y. Investigation of the performance of a stand-alone horizontal axis tidal current turbine based on in situ experiment. *Ocean. Eng.* **2016**, *113*, 111–120. [CrossRef]
16. Polinder, H.; Van der Pijl, F.F.; De Vilder, G.J.; Tavner, P.J. Comparison of direct-drive and geared generator concepts for wind turbines. *IEEE Trans. Energy Convers.* **2006**, *21*, 725–733. [CrossRef]
17. Hart, K.; McDonald, A.; Polinder, H.; Corr, E.J.; Carroll, J. Improved cost energy comparison of permanent magnet generators for large offshore wind turbines. In Proceedings of the European Wind Energy Association 2014 Annual Conference, Barcelona, Spain, 10–13 March 2014.
18. Djebarri, S.; Charpentier, J.F.; Scuiller, F.; Benbouzid, M. Design and performance analysis of double stator axial flux PM generator for rim driven marine current turbines. *IEEE J. Ocean. Eng.* **2015**, *41*, 50–66.
19. Titah-Benbouzid, H.; Benbouzid, M. Biofouling issue on marine renewable energy converters: A state of the art review on impacts and prevention. *Int. J. Energy Convers.* **2017**, *5*, 67–78. [CrossRef]
20. Mycek, P.; Gaurier, B.; Germain, G.; Pinon, G.; Rivoalen, E. Experimental study of the turbulence intensity effects on marine current turbines behaviour. Part I: One single turbine. *Renew. Energy* **2014**, *66*, 729–746. [CrossRef]
21. SHOM (Service Hydrographique et Ocanographique de la Marine), 3D Marine Tidal Currents in Fromveur (Ouessant island). 2014. Available online: https://diffusion.shom.fr/pro (accessed on 6 June 2019).
22. El Tawil, T.; Charpentier, J.F.; Benbouzid, M. Tidal energy site characterization for marine turbine optimal installation: Case of the Ouessant Island in France. *Int. J. Mar. Energy* **2017**, *18*, 57–64. [CrossRef]
23. Radzevich, S.P. *Dudley's Handbook of Practical Gear Design and Manufacture*; CRC Press: Boca Raton, FL, USA, 2012.
24. Guo, Y.; Parsons, T.; King, R.; Dykes, K.; Veers, P. *Analytical Formulation for Sizing and Estimating the Dimensions and Weight of Wind Turbine Hub and Drivetrain Components*; Technical Report; National Renewable Energy Lab. (NREL): Golden, CO, USA, 2015.
25. Harrison, R.; Hau, E.; Snel, H. *Large Wind Turbines: Design and Economics*; Wiley: Chichester, UK, 2000; Volume 1.
26. Rourke, F.O.; Boyle, F.; Reynolds, A. Marine current energy devices: Current status and possible future applications in Ireland. *Renew. Sustain. Energy Rev.* **2010**, *14*, 1026–1036. [CrossRef]
27. Dubois, M.R. Review of electromechanical conversion in wind turbines. *Rep. EPP00* **2000**, *3*, 4–10.
28. Djebarri, S. Contribution à la Modelisation et à la Conception Optimale de Generatrices à Aimants Permanents Pour Hydroliennes. Ph.D. Thesis, Université de Bretagne Occidentale, Brest, France, 2015.

29. Djebarri, S.; Charpentier, J.F.; Scuiller, F.; Benbouzid, M.; Guemard, S. Rough design of a double-stator axial flux permanent magnet generator for a rim-driven marine current turbine. In Proceedings of the 2012 IEEE International Symposium on Industrial Electronics, Hangzhou, China, 28–31 May 2012; pp. 1450–1455.
30. Pyrhonen, J.; Jokinen, T.; Hrabovcova, V. *Design of Rotating Electrical Machines*; John Wiley & Sons: Chichester, UK, 2013.
31. Böhmeke, G. Development and Operational Experience of the Wind Energy Converter WWD-1. In Proceedings of the 2003 Europ Wind Energy Conference, Madrid, Spain, 16–19 June 2003.
32. Boldea, I. *The Electric Generators Handbook-2 Volume Set*; CRC Press: Boca Raton, FL, USA, 2005.
33. Grauers, A. Efficiency of three wind energy generator systems. *IEEE Trans. Energy Convers.* **1996**, *11*, 650–657. [CrossRef]
34. Xue, Y.S.; Han, L.; Li, H.; Xie, L.D. Optimal design and comparison of different PM synchronous generator systems for wind turbines. In Proceedings of the 2008 International Conference on Electrical Machines and Systems, Wuhan, China, 17–20 October 2008; pp. 2448–2453.

© 2020 by the authors. Licensee MDPI, Basel, Switzerland. This article is an open access article distributed under the terms and conditions of the Creative Commons Attribution (CC BY) license (http://creativecommons.org/licenses/by/4.0/).

Article

A Synchronous Sampling Based Harmonic Analysis Strategy for Marine Current Turbine Monitoring System under Strong Interference Conditions

Milu Zhang [1,*], Tianzhen Wang [1,*], Tianhao Tang [1], Zhuo Liu [1] and Christophe Claramunt [1,2]

[1] Department of Electrical Automation, Shanghai Maritime University, No.1550 Haigang Avenue, Shanghai 201306, China; thtang@shmtu.edu.cn (T.T.); liuzhuo917@163.com (Z.L.); claramunt@ecole-navale.fr (C.C.)
[2] Naval Academy Research Institute, 29240 Brest, France
* Correspondence: mlzhang@shmtu.edu.cn (M.Z.); tzwang@shmtu.edu.cn (T.W.)

Received: 15 April 2019; Accepted: 27 May 2019; Published: 3 June 2019

Abstract: Affected by high density, non-uniform, and unstructured seawater environment, fault detection of Marine Current Turbine (MCT) faces various fault features and strong interferences. To solve these problems, a harmonic analysis strategy based on zero-crossing estimation and Empirical Mode Decomposition (EMD) filter banks is proposed. First, the detection problems of rotor imbalance fault under strong interference conditions are described through an analysis of the fault mechanism and operation environment of MCT. Therefore, against various fault features, a zero-crossing estimation is proposed to calculate instantaneous frequency. Last, and in order to solve the problem that the frequency and amplitude of the operating parameters are partially or completely covered by interference, a band-pass filter based on EMD is used, together with a characteristic frequency selected by a Pearson correlation coefficient. This strategy can accurately detect the multiplicative faults under strong interference conditions, and can be applied to the MCT fault detection system. Theoretical and experimental results verify the effectiveness of the proposed strategy.

Keywords: marine current turbine; multiplicative fault detection; zero-crossing estimation; EMD-based filter bank

1. Introduction

Timely maintenance can prolong the life of an MCT. Condition-based maintenance technology is a pivotal solution to reduce the long-term operation and maintenance cost of MCTs [1]. However, due to the complex marine environment, the condition-based maintenance of MCT faces the following problems: few monitoring variables, various fault features, and strong interferences [2]. In order to improve the safety and reliability of MCT, a fault detection strategy for non-stationary signals with strong interference is needed [3].

Condition monitoring provides continuous indications of components based on techniques, including vibration analysis, temperature, and acoustic emission analysis, among others [3]. Because of the limitations of underwater environment sensors, fault detection based on stator current has several advantages, since it is a non-invasive technique and avoids the use of additional sensors [4]. Regarding current-based detection methods, fault causes can be divided into two types: additive and multiplicative faults. Additive faults refer to the superposition of a pulse signal, a constant signal or a random small signal on the original normal signal [5], while multiplicative faults refer to signals with a certain frequency superimposed on the original normal signal [6]. For additive fault detection, time-domain statistical analysis can be used. A recursive principal component method can be applied to update the control limit adaptively to detect motor faults under unstable working conditions [7]. Multiplicative faults should be rather detected on the basis of energy frequency characteristics. Under stable working conditions, detections can be completed by frequency domain analysis. However, the magnitude of different frequency components changes with working conditions in practice [8]. Therefore, frequency domain analysis can be used in such conditions. Local feature detection, such as Short-Time Fourier Transforms (STFT) [9], Wavelet Transforms (WT) [10], and Hilbert-Huang transforms [11], can then be applied. STFT is used in [12] for blade crack detection of wind turbine to identify time-varying fault frequencies. In [13], motor current signature analysis is applied based on two sets of over complete wavelets; the gear fault-related component is highlighted under different operating conditions. In [14], SVD filtering combines the parameter estimation technique of rotation invariant signal to detect broken rotor bars under variable operating conditions. However, these methods identify the characteristic frequency, which amplitudes are very low in practice, and therefore may be partially or completely covered by the fundamental frequency and strong interference.

In order to solve these problems, we introduce a harmonic analysis strategy based on zero-crossing estimation and EMD filter banks. First, the detection problems of rotor imbalance fault under strong interference conditions are described through an analysis of the fault mechanism and operation environment of MCT. Therefore, against various fault features, zero-crossing estimation is proposed to calculate the instantaneous frequency. In order to solve the problem that the frequency and amplitude of the operating parameters are partially or completely covered by interference, a band-pass filter based on EMD is applied. Next characteristic frequency is selected by a Pearson correlation coefficient. Overall, this method can improve the detection accuracy of multiplicative faults, and has practical significance for ensuring the safe operation of MCT. The rest of the paper is organized as follows. Section 2 introduces the problem description, while Section 3 develops a harmonic analysis strategy. Section 4 presents some simulation results and analysis. Section 5 presents some experimental results and analysis, while Section 6 concludes the paper and draws perspectives for further work.

2. Problem Description

A MCT system harnesses energy from tidal flow, which converts kinetic energy into the motion of a turbine and then drives electrical generators. In comparison with wind turbines, MCTs have high energy density and small moment of inertia. Influenced by the tide, long-term flow velocity varies in a wide range, and short-term flow velocity varies frequently [15]. Waves, turbulence, and watershed topography are main interference factors affecting MCT working conditions [16]. For MCT multiplication faults, taking blade imbalance fault as an example, specific frequencies can be detected in stator current. However, affected by interference, the monitoring signal has multiple time scales and multiple characteristics [17]. This seriously affects the accuracy of fault detection.

2.1. Multiplication Fault of MCT

This section mainly describes the multiplicative faults of MCT under strong interference. Take blade imbalance fault as an example; this fault refers to the signal with a certain 1P frequency superimposed on the original normal stator current signal.

The kinetic energy harnessed by the turbine can be described as:

$$C_p \rho A V_{current}^3 / 2 = T_{mech} \omega_m \quad (1)$$

where C_p is power coefficient, ρ the water density, A the cross-sectional area of the turbine, $V_{current}$ the tidal current velocity, T_{mech} and ω_m is mechanical torque and speed. The turbine drives an electrical generator. Considering the generator as a permanent magnet synchronous motor, the motion equation is given as:

$$J_m d\omega_m / dt = T_{mech} - T_e - f_v \omega_m \quad (2)$$

where J_m is the moment of inertia, f_v friction coefficient, T_e electromagnetic torque. The current output can described as:

$$i_s = I_s \cos(\int_{t_0}^{t} p\omega_m d\tau) \quad (3)$$

where I_s is the amplitude of the stator current, p the number of pole pairs. The relationship between these variables is shown in Figure 1.

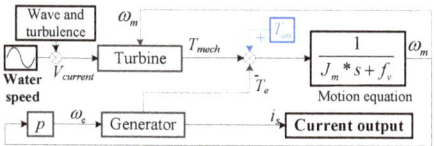

Figure 1. Energy transfer mechanism of MCT.

When imbalance fault happens, additional torque appears as shown in Figure 1. Mechanical speed accelerates when imbalance mass falls in the direction of rotation, and mechanical speed decelerates when imbalance mass goes upward in the direction of reverse rotation, as shown in Figure 2.

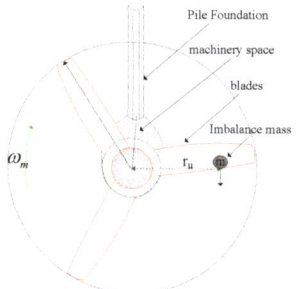

Figure 2. Effect of imbalance mass on the MCT.

Additional torque caused by imbalance fault can be described as:

$$T_{im}(t) = F_g r_u \cdot \sin(\omega_m t) \quad (4)$$

where F_g is resultant of forces downward, r_u is the distance between attachments to the center of the shaft. Bring additional torque T_{im} into Equation (2), then additional speed can be obtained as:

$$\Delta \omega_m = F_g r_u \cdot \cos(\omega_m t) / J_m \omega_m = 2\pi B \cdot \cos(\omega_m t) \quad (5)$$

Considering $\Delta \omega_m \ll \omega_m$, the stator current can be expressed as:

$$i_s(t) = I_s \cdot \cos[p(\omega_m + \Delta \omega_m)t + \gamma] \quad (6)$$

where γ is the initial angle. In this case, stator current frequency is $f_s = (\omega_m + \Delta \omega_m)/(2\pi \cdot p)$, the fault feature frequency is $f_{im} = \Delta \omega_m / (2\pi \cdot p)$.

2.2. The Effect of Strong Interference

With a changeable water speed, mechanical torque T_{mech} can be written as:

$$T_{mech} = T_{sl} + T_{fa} \quad (7)$$

where T_{sl} stands for slowly time-varying torque, T_{fa} stands for abrupt change torque. The torque T_{sl} is caused by long-term tidal velocity, as shown in Figure 3. It provides one month water flow velocity variation curve—the velocity changes in a large range. The torque T_{fa} is caused by spatial velocity change, as shown in Figure 4. T_{fa} changes frequently due to unstable flow velocity, which results in strong interference.

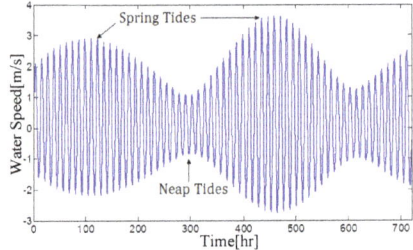

Figure 3. Long-term tidal velocity.

(a) Stable flow velocity (b) Unstable flow velocity

Figure 4. Spatial velocity change.

Combining Equations (7) and (2), mechanical speed can be rewritten as:

$$\omega_m = \omega_{sl} + \omega_{fa} \tag{8}$$

Accordingly, $\omega_{sl} = 2\pi f_{sl}$ stands for slowly time-varying rotating speed. $\omega_{fa} = 2\pi f_{fa}$ stands for abrupt change rotating speed. The strong interference factor is taken into the calculation of stator current in Equation (6), which can be rewritten as:

$$\begin{aligned} i_s(t) &= I_s \cdot \cos[p(\omega_{sl} + \omega_{fa} + \Delta\omega_m)t + \gamma] \\ &= I_s \cdot \cos[2\pi p(f_{sl} + f_{fa} + \Delta f_m)t + \gamma] \end{aligned} \tag{9}$$

The fault feature frequency f_{im} changes with the mechanical cycle; f_{im} may be partially or completely covered by interference frequency f_{fa}.

Figure 5 shows the mechanical phases in three cases: constant rotating speed, slowly time-varying rotating speed, and abrupt change rotating speed. In Figure 5a, imbalance faults can be accurately distinguished. However, when the rotating speed changes slowly, the imbalance fault feature frequency changes as shown in Figure 5b. When the rotating speed changes frequently, imbalance fault are not able to detect, features are partially or completely covered by interference, as shown in Figure 5c.

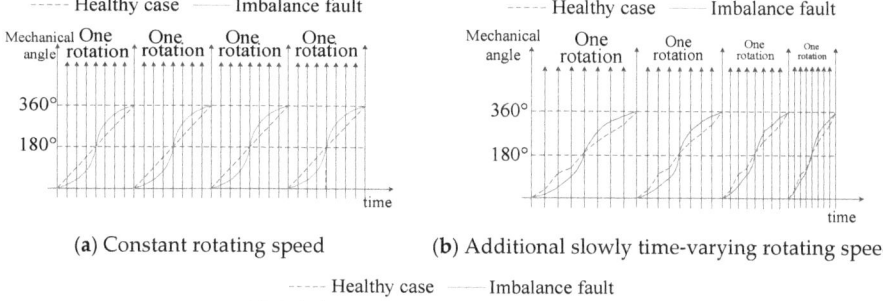

(a) Constant rotating speed

(b) Additional slowly time-varying rotating speed

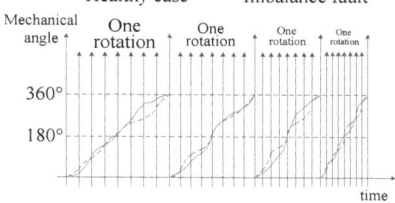

(c) Additional abrupt change rotating speed

Figure 5. Mechanical phase affected by interference.

3. Proposed Harmonic Analysis Strategy

This section aims to solve the fault feature extraction problem under the condition of variable speed operation and strong disturbance. Fault feature has many different possible forms under strong interference conditions. The proposed detection method consists of three parts: The zero-crossing estimation is used to unify characteristic frequency; the interference signal is decomposed to several intrinsic mode functions by EMD filter banks; the specific data for extracting the characteristic frequency is selected by Pearson correlation coefficient.

3.1. Instantanous Frequency Caculation by Zero-Crossing Estimation

The fault feature has many different possible forms. If the original stator current signal is uniformly sampled with sampling frequency, Equation (6) can be rewritten as:

$$i_s(n) = I_s(n) \cdot \cos[\varphi(n)] \quad (10)$$

Find all zero-crossing pairs $P(i) = \{i_s(n_i), i_s(n_i+1)\}$. If $i_s(n_i) > 0(< 0)$ then $i_s(n_i+1) < 0(> 0)$. Using linear interpolation for each zero-crossing point of the sequence, we can get the time value of the zero-crossing point:

$$t_{zero} = [i_s(n_i)t_{ni+1} - i_s(n_i+1)t_{ni}] / [i_s(n_i) - i_s(n_i+1)] \quad (11)$$

where t_{ni} is the time of $i_s(n_i)$, t_{ni+1} is the time of $i_s(n_i+1)$. The total number of sampling points in each mechanical cycle is variable because of the changing shaft rotating speed. Assume $N_0, N_k, \ldots, N_z$ are the zero-crossing points for one mechanical cycle, $k = 0, 1, \ldots, Z$ (Z is determined by the number of generator poles). The instantaneous frequency defined at N_k is:

$$f_{N_k} = (f_{N_k - N_{k-1}} + f_{N_{k+1} - N_k} + f_{N_{k+1} - N_{k-1}})/5 \quad (12)$$

where $f_{N_k - N_{k-1}} = 1/[t_{zero}(k) - t_{zero}(k-1)]$, $f_{N_{k+1} - N_k} = 1/[t_{zero}(k+1) - t_{zero}(k)]$, $f_{N_{k+1} - N_{k-1}} = 1/[t_{zero}(k+1) - t_{zero}(k-1)]$. Then, cubic spline interpolation is used to obtain the instantaneous frequency curve.

For the mechanical cycle, the time span is N_0 to N_Z. Each mechanical cycle is different taking into account the effect of rotational speeds. The average time of mechanical cycle can be denoted as $\bar{t}$. There are two major parts to unify the fault feature form:

1). Scaling on the Time Axis

In order to unify fault feature frequency, set up the total time of each mechanical cycle $t_{zero}(Z) = \bar{t}$, then the time for each zero-crossing point is changed to:

$$t_{zero}(k) = \bar{t} \cdot k / Z \tag{13}$$

By Equation (13), we reconstruct a time series, and a unified characteristic frequency can be the obtained.

2). Eliminate the Trend Component

As shown in Figure 6, Least Square (LS) is used to eliminate the trend component according to the size of the windows. Different fluctuation curve is retained to highlight the fault degree. For 1P frequency, the length of window is selected as $t_{zero}(Z)$. After LS, the fluctuation curve of instantaneous frequency is obtained.

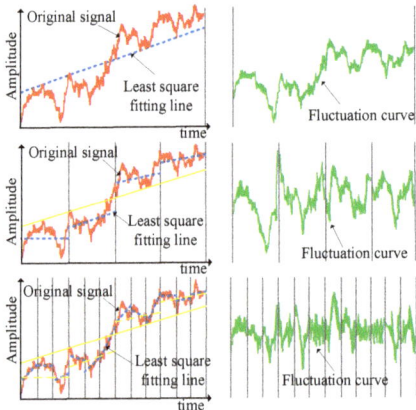

Figure 6. LS with different size of windows.

3.2. Interference Filtering by EMD Filter Banks

In order to effectively remove the influence of strong interference, an EMD-based filter bank is proposed. The aim of this method is to extract the characteristic frequency signal from the original dataset. Affected by interference, the amplitude of characteristic frequency cannot be accurately obtained. EMD-based filter bank decomposes the interference signal into several intrinsic mode functions to reduce the influence of interference on characteristic frequency signals. Sampling frequency F_s is then set up. From Equation (5), the frequency variation affected by interference can be obtained:

$$\Delta f_m(k) = B \cdot \cos\left(2\pi \left(f_{sl} + f_{fa}\right)k/F_s\right) \tag{14}$$

In consideration of $f_{sl} \gg f_{fa}$, $f_{fa} = \delta \approx 0$ can be approximated as zero; Equation (14) can be rewritten as:

$$\begin{aligned}\Delta f_m(k) &= B\cos(2\pi k f_{sl}/F_s)\cos(2\pi k\delta/F_s) \\ &\quad -B\sin(2\pi k f_{sl}/F_s)\sin(2\pi k\delta/F_s) \\ &\approx B\cos(2\pi k f_{sl}/F_s)\cdot 1 + \xi(\delta)\end{aligned} \tag{15}$$

where $\xi(\delta)$ is the noise component caused by strong interference. Equation (15) is expressed as the sum of characteristic frequency components and noise components, which can be composed of a set of intrinsic mode functions (IMFs) and residual terms:

$$\Delta f_m(k) = \sum_{i=1}^{I} C_i(k) + r(k) \tag{16}$$

where $i = 1, 2, \ldots I$, $r(k)$ is residual term, $C_i(k)$ is used to indicate IMFs, can be represented as:

$$C_i(k) = r_i(k) - \sum_{j=1}^{J_i} m_{i,j}(k) \tag{17}$$

where $m_{i,j}(k)$ is the average value of the upper envelope and the lower envelope, J_i is maximum number of iterations to calculate the intrinsic mode function. Average trend of the signal is:

$$r_i(k) = \begin{cases} \Delta f_m(k) & i = 1 \\ \sum_{j=1}^{J_{i-1}} m_{i-1,j}(k) & i = 2, 3, \ldots I \end{cases} \tag{18}$$

If there is no interference, $\xi(\delta) = 0$. The average value of the upper envelope and the lower envelope $m_{i,j}(k)$ become zero. Fault feature frequency is obtained in $C_i(k) = r_i(k) = \Delta f_m(k)$. EMD can make $m_{i,j}(k)$ tending to the null value; the interference components are distributed into different IMFs.

3.3. Characteristic Frequency Selection by Pearson Correlation Coefficient

The Pearson correlation coefficient is used to determine which IMF contains fault feature frequency. With $X(k)$ and $Y(k)$ two-time series, Pearson correlation coefficient is defined as:

$$R(X, Y) = \frac{\sum_k [X(k) - \overline{X}] \cdot [Y(k) - \overline{Y}]}{\sqrt{\sum_k [X(k) - \overline{X}]^2} \cdot \sqrt{\sum_k [Y(k) - \overline{Y}]^2}} \tag{19}$$

where $\overline{X}$ and $\overline{Y}$ is the average value. The value of $R(X, Y)$ is from -1 to 1; -1 indicates negative correlation, 1 indicates positive correlation, and 0 irrelevant. Assume that $X(k) = \Delta f_m(k)$ and $Y(k) = IMF_i(k)$, because only one IMF_i contains characteristic frequency and the other contains the noise component; the largest $R(X, Y)$ is selected to realize EMD filtering.

3.4. Harmonic Analysis Strategy for Fault Detection

The procedure of harmonic analysis strategy for multiplier fault under strong interference is shown in Figure 7. The details are as follows:

(1) Measuring the current signal from MCT.
(2) Zero-crossing estimation is used to calculate the instantaneous frequency by Equation (12). Then, a unified characteristic frequency is obtained by reconstructing the time series, and the LS method is used to eliminate the trend component. By combining historical data, the estimated instantaneous frequency is used to identify the work conditions of MCTs.
(3) To solve the problem that the frequency and amplitude of the operating parameters are partially or completely covered by interference, an EMD filter bank is used to remove interference in the frequency band.
(4) The IMF selected by Pearson correlation coefficient is analyzed by spectrum analysis, which provides fault feature frequency.

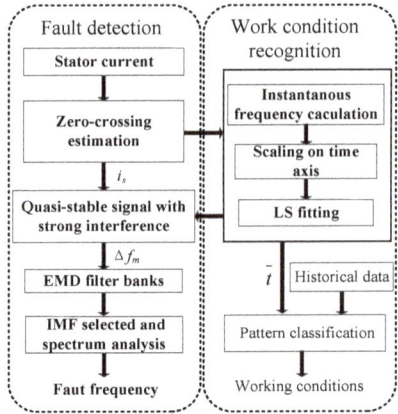

Figure 7. Flow chart of the proposed method.

4. Simulation Results and Analysis

A sample example is given to illustrate the effectiveness of the proposed method. An AM/FM signal is given as:

$$x(t) = (1 + 0.4\sin(2\pi \cdot 2t)) \cdot \cos[2\pi \cdot 16t - 0.8\cos(2\pi \cdot 2t) - 0.08\cos(2\pi \cdot 80t)]$$

Signal fundamental frequency is 16 Hz, FM frequency 2 Hz, high-order FM frequency 80 Hz, AM frequency 2 Hz. The instantaneous frequency of the modulated signal is then obtained:

$$\begin{aligned} f(t) &= d[2\pi \cdot 16t - 0.8\cos(2\pi \cdot 2t) - 0.008\cos(2\pi \cdot 80t)]/(2\pi \cdot dt) \\ &= 16 + 1.6\sin(2\pi \cdot 2t) + 0.64\sin(2\pi \cdot 80t) \end{aligned} \quad (20)$$

Figure 8 shows the time-frequency spectrum used by Hilbert transform and separately proposed the zero-crossing estimation method. From Figure 8a, Hilbert transform can present high-order harmonics. However, the boundary effects affect estimation accuracy. Figure 8b shows the FM frequency 2 Hz more clearly and gets rid of the high frequency harmonics by scaling the time axis and Equation (13). The time span in Figure 8a is different from Figure 8b due to time axis expansion and contraction. In Equation (12), $\bar{t} = 0.5$, $Z = 8$. The proposed zero-crossing estimation method realizes purposeful fault feature frequency extraction, and overcomes shortcomings of the Hilbert transform.

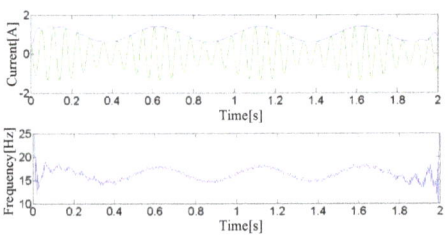

(a) Based on Hilbert transform

Figure 8. Cont.

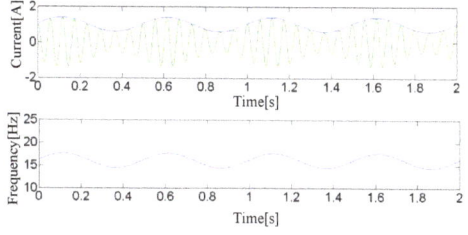

(**b**) Based on zero-crossing estimation

Figure 8. Simulation test result of time-frequency spectrum.

5. Experimental Results and Analysis

5.1. MCT Experiment and Analysis

In order to verify the proposed method, a MCT prototype is implemented in an experimental platform, as shown in Figure 9. This experiment platform consists of the following parts: (1) current simulation system (enclosed water channel: adjustable flow velocity 0.2m/s-1.5m/s); (2) MCT prototype (PMSG: 8 pole-pairs); (3) data acquisition and monitoring system (sampling frequency 1 kHz); (4) imbalance fault setting, as shown in Figure 10.

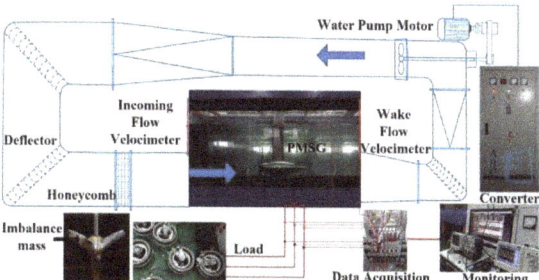

Figure 9. MCT experimental system.

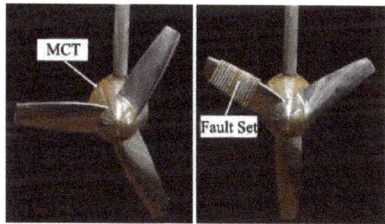

Figure 10. Imbalance fault setting.

Figure 11 shows the stator current of MCT under a strong interference working condition [18]. One can note that when the velocity changes frequently, the collected waveforms fluctuate sharply. Under the condition of abrupt change flow velocity, both the original signal and the instantaneous frequency vary frequently with the change of flow velocity. The fault feature frequency is partially or completely covered by interference. The experimental phenomena coincide with the theoretical analysis, as shown in Equation (9).

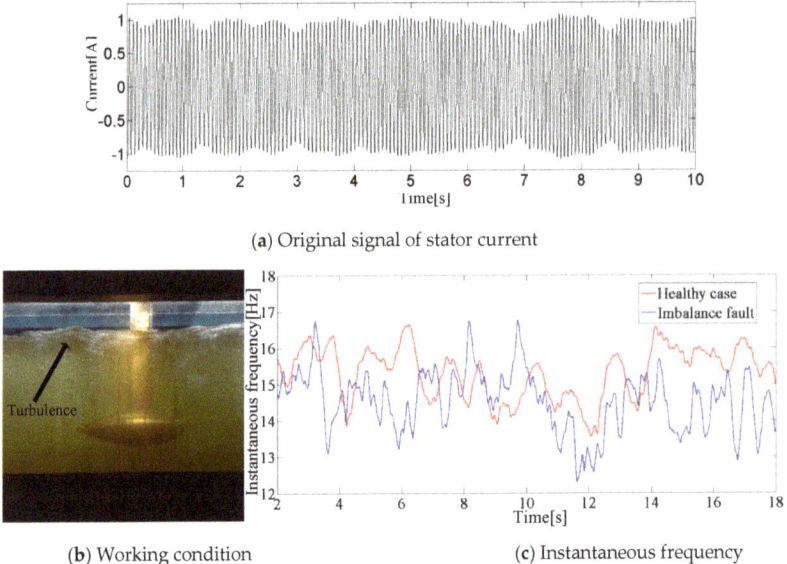

(a) Original signal of stator current

(b) Working condition

(c) Instantaneous frequency

Figure 11. MCT strong interference working condition.

5.2. Fault Detection and Analysis

The proposed method is used to detect faults in two cases: the normal operation of the MCT and 3% imbalanced faults. The magnitude ratio of the imbalanced torque to the total mechanical torque of the MCT is used to describe the degree of the faults.

Figure 12 shows the detection results after an EMD-based filter bank. The original signal is decomposed into IMF1-IMF8. Noise or interference with similar frequencies is distributed to the same IMF, which can highlight fault feature frequency. After separating the interference component, the amplitude of characteristic frequency obtained is more accurate; this can help identify the fault degree. In Figure 12a, the maximum value of $R(X,Y)$ is 0.104 to IMF2. In Figure 12b, the maximum value of $R(X,Y)$ is 0.973 to IMF2.

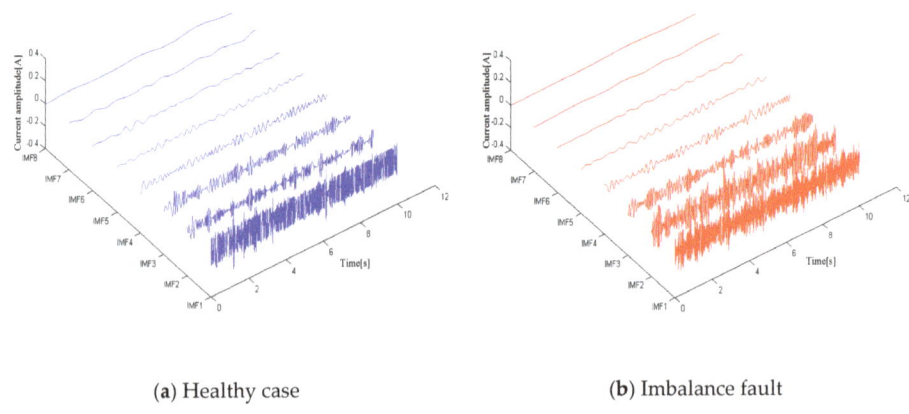

(a) Healthy case

(b) Imbalance fault

Figure 12. Decomposition results of an EMD-based filter bank.

In Figure 13, it can be seen that the spectrum appears near the frequency of 1.875 Hz in IMF2, which corresponds to the characteristic frequency (1P frequency) of the imbalanced fault of MCT system.

Several frequency components generated by strong interference are distributed on different IMFs by EMD, which greatly reduces the influence of interference and highlights the fault feature frequency.

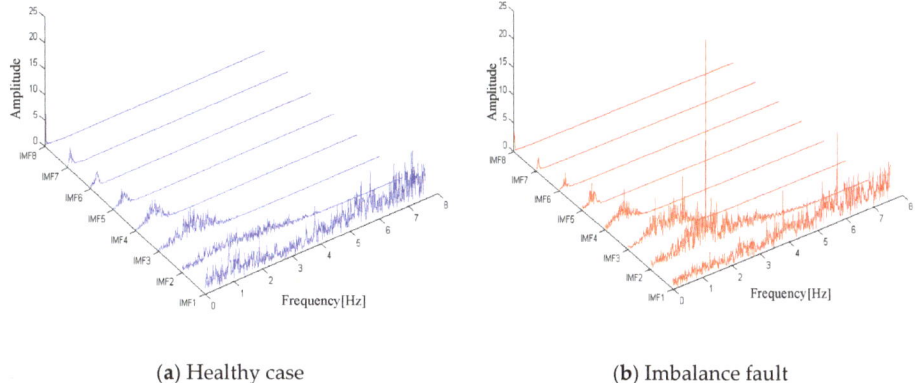

(a) Healthy case   (b) Imbalance fault

Figure 13. Proposed detection result under strong interference conditions.

Figure 14 is the front panel of the block diagram program designed by LABVIEW. The proposed strategy is used in this monitoring system. The detection result of the MCT system is given, including short-term original signal, time-frequency spectrum, frequency difference by LS method and fault feature frequency. The last is the spectrum diagram of IMF, which has maximum value of $R(X, Y)$. The input signal of the monitoring system is the stator current, and the output of the system is the working condition of the MCT and fault condition. The experimental results show that the proposed method can effectively eliminate strong interference and solve the problem that the frequency and amplitude of the operating parameters are partially or completely covered by interference.

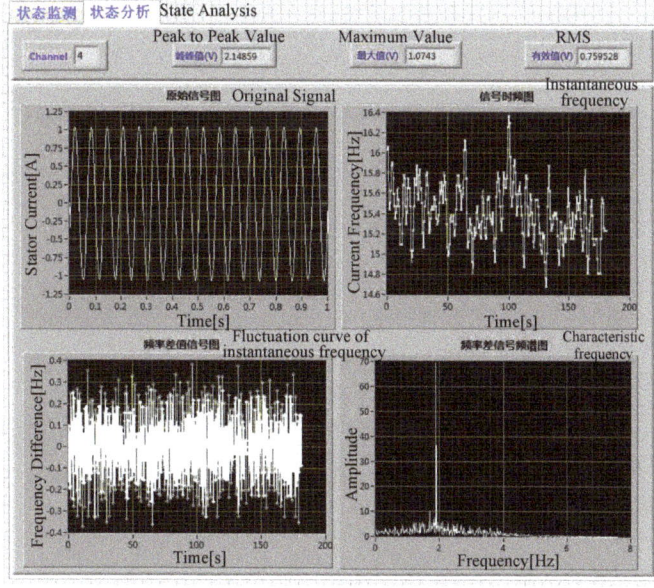

Figure 14. Detection results of the monitoring system.

6. Conclusions

A MCT fault detection method based on zero-crossing estimation and EMD-based filter bank is proposed to deal with strong interference conditions. This method can identify working conditions by scaling on time axis and the LS fitting method. EMD is used to realize the function of filter banks by distributing the interference frequency in different IMFs. Last, Pearson correlation coefficient is used to determine which IMF contains fault characteristics frequency. The following conclusions are drawn:

(1) Zero-crossing estimation can identify specific frequencies purposefully, and effectively identify working conditions. On combining with EMD-based filter banks, the proposed method can effectively solve the problem that the frequency and amplitude of operating parameters are partially or completely covered under strong interference.

(2) Zero-crossing estimation can be easily implemented in the monitoring system, compared to STFT and WT. This method has good performance to estimate instantaneous frequency and can be used for long-term monitoring of current machine systems. However, this method has a low frequency resolution, not suitable for short-time signal processing.

It is worth pointing out that the real marine environment is more disturbing than what we simulate [19]. Due to wave, turbulence, and marine organisms, monitoring signals exhibit unstable, non-linear and low signal-to-noise ratio characteristics [20]. The proposed method may be more needed in these harsh environments.

Author Contributions: Conceptualization, M.Z.; methodology, M.Z.; software, Z.L.; validation, M.Z.; formal analysis, M.Z.; investigation, M.Z., Z.L.; resources, M.Z., T.T.; writing—original draft preparation, M.Z.; writing—review and editing, C.C.; visualization, C.C.; supervision, T.T.; project administration, T.T.; funding acquisition, T.W.

Funding: This paper is supported by the National Natural Science Foundation of China (61673260) and Shanghai Natural Science Foundation (16ZR1414300).

Conflicts of Interest: The authors declare no conflict of interest.

References

1. Besnard, F.; Bertling, L. An Approach for Condition-Based Maintenance Optimization Applied to Wind Turbine Blades. *IEEE Trans. Sustain. Energy* **2010**, *1*, 77–83. [CrossRef]
2. Zhang, M.; Wang, T.; Tang, T.; Benbouzid, M.; Diallo, D. An imbalance fault detection method based on data normalization and EMD for marine current turbines. *ISA Trans.* **2017**, *68*, 302–312. [CrossRef]
3. Márquez, F.P.G.; Tobias, A.M.; Pérez, J.M.P.; Papaelias, M. Condition monitoring of wind turbines: Techniques and methods. *Renew. Energy* **2012**, *4*, 169–178. [CrossRef]
4. Feng, Z.; Chen, X.; Zuo, M.J. Induction Motor Stator Current AM-FM Model and Demodulation Analysis for Planetary Gearbox Fault Diagnosis. *IEEE Trans. Ind. Inform.* **2018**, *15*, 2386–2394. [CrossRef]
5. Tellili, A.; ElGhoul, A.; Abdelkrim, M.N. Additive fault tolerant control of nonlinear singularly perturbed systems against actuator fault. *J. Electr. Eng.* **2017**, *68*, 68–73. [CrossRef]
6. Tan, C.P.; Edwards, C. Multiplicative fault reconstruction using sliding mode observers. In Proceedings of the 2004 5th Asian Control Conference, Melbourne, Australia, 20–23 July 2004; Volume 2, pp. 957–962.
7. Yu, T.; Wang, X.; Shami, A. Recursive Principal Component Analysis-Based Data Outlier Detection and Sensor Data Aggregation in IoT Systems. *IEEE Internet Things J.* **2017**, *4*, 2207–2216. [CrossRef]
8. Elghali, S.E.B.; Benbouzid, M.E.H.; Charpentier, J.F. Marine Tidal Current Electric Power Generation Technology: State of the Art and Current Status. In Proceedings of the 2007 IEEE International Electric Machines & Drives Conference, Antalya, Turkey, 3–5 May 2007; Volume 2, pp. 1407–1412.
9. Satpathi, K.; Yeap, Y.M.; Ukil, A.; Geddada, N. Short-Time Fourier Transform Based Transient Analysis of VSC Interfaced Point-to-Point DC System. *IEEE Trans. Ind. Electron.* **2018**, *65*, 4080–4091. [CrossRef]
10. Zhang, L.; Lang, Z.-Q. Wavelet Energy Transmissibility Function and Its Application to Wind Turbine Bearing Condition Monitoring. *IEEE Trans. Sustain. Energy* **2018**, *9*, 1833–1843. [CrossRef]
11. Koganezawa, S. Frequency analysis of disturbance torque exerted on a carriage arm in hard disk drives using Hilbert-Huang Transform. *IEEE Trans. Magn.* **2018**, *99*, 1–6. [CrossRef]

12. Jian-Zhong, W.U.; Yi, T. STFT-based crack detection on wind turbine blades. *Chin. J. Constr. Mach.* **2014**, *12*, 180–183.
13. Chai, N.; Yang, M.; Ni, Q.; Xu, D. Gear fault diagnosis based on dual parameter optimized resonance-based sparse signal decomposition of motor current. *IEEE Trans. Ind. Appl.* **2018**, *54*, 3782–3792. [CrossRef]
14. Brenner, M.J. Non-Stationary Dynamics Data Analysis with Wavelet-Svd Filtering. *Mech. Syst. Signal Process.* **2001**, *17*, 765–786. [CrossRef]
15. Zhang, M.; Wang, T.; Tang, T.; Benbouzid, M.; Diallo, D. Imbalance fault detection of marine current turbine under condition of wave and turbulence. In Proceedings of the IECON 2016—42nd Annual Conference of the IEEE Industrial Electronics Society, Florence, Italy, 24–27 October 2016; pp. 6353–6358.
16. Zhang, M.; Wang, T.; Tang, T.; Wang, Y. Blade Imbalance Fault Detection Method for Direct-Driven Marine Current Turbine with Permanent Magnet Synchronous Generator. *Trans. China Electrotech. Soc.* **2018**, *33*, 38–47.
17. Samantaray, L.; Dash, M.; Panda, R. A Review on Time-frequency, Time-scale and Scale-frequency Domain Signal Analysis. *IETE J. Res.* **2005**, *51*, 287–293. [CrossRef]
18. Lust, E.E.; Luznik, L.; Flack, K.A.; Walker, J.M.; Van Benthem, M.C. The influence of surface gravity waves on marine current turbine performance. *Int. J. Mar. Energy* **2013**, *3*, 27–40. [CrossRef]
19. Thomson, J.; Polagye, B.; Durgesh, V.; Richmond, M.C. Measurements of Turbulence at Two Tidal Energy Sites in Puget Sound, WA. *IEEE J. Ocean. Eng.* **2012**, *37*, 363–374. [CrossRef]
20. Keenan, G.; Sparling, C.; Williams, H.; Fortune, F.; Davison, A. *SeaGen Environmental Monitoring Programme: Final Report*; Royal Haskoning Enhancing Society: Amersfoort, The Netherlands, 2011.

© 2019 by the authors. Licensee MDPI, Basel, Switzerland. This article is an open access article distributed under the terms and conditions of the Creative Commons Attribution (CC BY) license (http://creativecommons.org/licenses/by/4.0/).

Article

Analysis of a Horizontal-Axis Tidal Turbine Performance in the Presence of Regular and Irregular Waves Using Two Control Strategies

Stephanie Ordonez-Sanchez [1], Matthew Allmark [2,*], Kate Porter [1], Robert Ellis [2], Catherine Lloyd [2], Ivan Santic [3], Tim O'Doherty [2] and Cameron Johnstone [1]

1. Energy Systems Research Unit, University of Strathclyde, Glasgow G1 1XJ, UK; s.ordonez@strath.ac.uk (S.O.-S.); kate.porter.10@ucl.ac.uk (K.P.); cameron.johnstone@strath.ac.uk (C.J.)
2. School of Engineering, Cardiff University, Queen's Buildings, The Parade, Cardiff CF24 3AA, UK; ellisR10@cf.ac.uk (R.E.); lloydC11@cf.ac.uk (C.L.); odoherty@cf.ac.uk (T.O.)
3. CNR-INM, Consiglio Nazionale delle Ricerche, Istituto di Ingegneria del Mare, 00128 Rome, Italy; ivan.santic@insean.cnr.it
* Correspondence: allmarkmj1@cf.ac.uk; Tel.:+44-2920-8759-05

Received: 19 December 2018; Accepted: 21 January 2019; Published: 24 January 2019

Abstract: The flow developed on a tidal site can be characterized by combinations of turbulence, shear flows, and waves. Horizontal-axis tidal turbines are therefore subjected to dynamic loadings that may compromise the working life of the rotor and drive train components. To this end, a series of experiments were carried out using a 0.9 m horizontal-axis tidal turbine in a tow tank facility. The experiments included two types of regular waveforms, one of them simulating an extreme wave case, the other simulating a more moderate wave case. The second regular wave was designed to match the peak period and significant wave height of an irregular wave which was also tested. Measurements of torque, thrust, and blade-bending moments were taken during the testing campaign. Speed and torque control strategies were implemented for a range of operational points to investigate the influence that a control mode had in the performance of a tidal stream turbine. The results showed similar average power and thrust values were not affected by the control strategy, nor the influence of either the regular or irregular wave cases. However, it was observed that using torque control resulted in an increase of thrust and blade root bending moment fluctuations per wave period. The increase in fluctuations was in the order of 40% when compared to the speed control cases.

Keywords: control strategy; dynamic loading; horizontal-axis tidal turbine; regular waves; irregular waves; tow tank

---

1. Introduction

Achieving commercial and financial viability in the marine energy industry is challenging due to the complex and variant nature of the conditions seen by Horizontal-Axis Tidal Turbines (HATTs). Non-uniformity and unsteadiness occurring in the marine environment in the form of turbulence, sheared flows, surface waves, and tidal cycles mean that tidal turbines are subjected to a wide range of dynamic loading characteristics. The turbine components must be able to withstand the maximum forces induced by the hydrodynamics and be resistant to fatigue damage due to the cyclic nature of these loads. The high variability in the flow characteristics also has implications for optimization of the device in terms of the sizing of components. Fluctuations in power production, resultant from the dynamic tidal resource, pose significant challenges for the power conditioning and control systems. Therefore, it is imperative that the realistic loading characteristics are quantified and the way in which these interact with the turbine system are fully understood.

The potential impacts of wave-current flows on the performance of HATTs have been highlighted by several authors through the use of numerical models. Reference [1] conducted a sensitivity analysis on a Blade Element Momentum Theory (BEMT) model and showed that wave height and spatial non-uniformity in the current are among the most influential parameters on the loading of a turbine rotor. The eccentric thrust loads generated are transmitted through the drivetrain, directly affecting the internal components (e.g., bearings and seals), as later demonstrated in [2]. Reference [3] used computational fluid dynamics techniques (CFD) to investigate the effects of oscillatory motion of waves on HATTs blades. They found that combined wave-current conditions have a substantial influence on the bending moments acting on the turbine, which translates into damaging effects on the drivetrain components. Similarly, Reference [4] employed a CFD model to demonstrate that the maximum thrust force on a tidal stream rotor in wave-current flow increased by approximately 16% compared to that under the relevant current alone case. They also concluded that dynamic stall can occur in some conditions, depending on the turbine position relative to the wave phase.

A handful of experimental studies to support the computational work analyzing the effects of wave-current interactions on small-scale tidal turbine have also been conducted by [5–8]. A summary of the test parameters used in these investigations is presented in Table 1. Many of these studies analyzed the turbine loads induced by a single waveform (one wave period and wave height combination), and those that included variation in wave height, wave period, and current velocity have limited the tests to only a small range of values. Conditions that would be more consistent with extreme waves have not been included in the studies. In some of the studies the results are also limited to a small portion of the power curve (i.e., small variation in the angular velocity of the rotor) and thus the wave-current effects are not fully visualized. The high blockage ratio present in the study of [5] will have significantly influenced the performance of the turbine, as explained in [9,10], hindering comparison of their results with the other studies.

The general consensus from these studies is that the addition of waves to a current does not significantly affect the average value of thrust or torque. However, the torque and thrust fluctuations are substantially increased with the addition of waves, for example [6] found that these were 2–3 times higher than under current alone. The reported size of these fluctuations does vary between the studies, indicating that better quantification of the interaction effects between the different test parameters is necessary to understand how the magnitude of the loading fluctuations changes over a full range of flow conditions.

A further consideration regarding the existing experimental studies is the influence of the facility used. The studies of [7,8] were conducted in tow tanks and consequently without the presence of turbulence. In contrast those of [5,6] were conducted in a flume with a turbulent current; the different turbulence intensities in their test programs are included in Table 1. The facility type will also affect the shape of the waveform, as the turbine is towed into the waves in the tow tank, but the waves travel with the current in the flume tests. The impact of these differences on the turbine loads is not well understood. Direct comparison of the different test campaigns in Table 1 would help elucidate some of these effects, but this is problematic due to the widely different parameter values used in each program.

Further testing programs related to the investigation of waves-current interactions and tidal stream turbines have also been investigated by [11–13]. Reference [11] carried out several experiments to investigate the effects of oblique waves in the loading of tidal turbines where it was observed that the presence of in-line waves was more detrimental to the tidal turbine than yawed waves. In a similar context, References [12,13] examined the structural and power variations in a tidal turbine prototype when this was subjected to opposing and focused waves. It was found that peak load and power outputs increased by up to 85% and 200%, respectively, when focused waves were used and were compared to current only conditions. This research also shows the necessity of quantifying the thrust loadings that a tidal turbine will need to withstand under extreme weather conditions.

Table 1. Summary of experiments undertaken for wave-curreny interactions with three-blade HATTs.

Author	Tank Type	Rotor Diameter m	Blockage Ratio %	Current/Tow Speed m s^{-1}	Turbulence Int.%	Wave Height m	Wave Period s
Barltrop et al. (2007) [8]	Tow	0.4	2.49	0–1.2	-	0.10	1.20
						0.10	1.20
						0.10	1.60
						0.10	1.70
						0.10	2.26
						0.02–0.14 *	1.20
Gaurier et al. (2013) [6]	Flume	0.9	8.8	0.67	5	0.16	2.00
				0.67		0.16	1.43
				0.68		0.28	1.43
Galloway et al. (2014) [7]	Tow	0.8	9.4	0.9	-	0.15	2.00
						0.10	2.00
Henriques et al. (2015) [5]	Flume	0.5	36.9	0.5	2	0.04	0.70
						0.08	0.90

* range of values.

It is worth mentioning that the prototypes employed by the authors cited in Table 1 and above have been designed to operate under a consistent control mode. According to the literature, the control mechanism used on those investigations was either based on controlling the torque or speed developed by the turbine with most of them operated under speed control. However, correlations between the control method used to achieve the desired operating region and structural loadings produced on the turbines were never discussed. This type of investigation is notably relevant as commercial tidal turbines usually work under a variable-speed controller, in a similar fashion as wind turbines [14].

Recent work published by [15] demonstrated the applicability of using an open-loop control strategy and a constant rotational speed proportional–integral–derivative (PID) feedback loop control on a 1.5 m diameter three-bladed horizontal-axis turbine. Their research showed that an increase in peak performance of almost 14% was achieved when maintaining a constant rotational speed in comparison to an open-loop control strategy. Similarly, Reference [16] explored the use of two reinforcement learning algorithms to control a tidal turbine based on maximum power point tracking. These simulated strategies included a Q-learning and a Neural Fitted Q-iteration, both were assessed in different flow conditions; i.e., current only, turbulence, and two types of waves. The outcomes showed that by using both strategies, the algorithms converged to the optimal power coefficient. However, there is no mention of how each control strategy reflected into the temporal fluctuations of the rotor loading.

To the authors' knowledge, the latter has only been initially examined by [17]. In that study, two tests were conducted on a small-scale turbine where the equivalent operating point was set with torque and speed control modes. It was observed that the torque-controlled test resulted in a slightly higher average blade force than with the corresponding speed-controlled test. However, this preliminary study only included a single test done in the region of peak power conditions. Therefore, the research presented here aims to expand on that initial work by investigating a wide range of operating conditions and three types of waveforms, including regular and irregular waves.

2. Methodology

The experimental campaign encompassed testing of a small-scale horizontal-axis turbine of 0.9 m rotor diameter at the CNR-INM (formerly INSEAN) towing tank in Rome. The tank dimensions are $9 \times 3.5 \times 220$ m, resulting in a blockage ratio of 2.8%. The low blockage characteristics in the facility ensured little influence of this should be seen in any of the tests. The center of the turbine hub was installed at 1.5 m below the still water surface on a steel stanchion clamped from above and positioned in the mid-section of the tank as shown in Figure 1.

Figure 1. Test setup in the CNR-INM wave-tow facility. Horizontal-axis turbine mounted on the tow carriage via stanchion with a set of ultrasonic wave probes on the left.

2.1. Turbine Design Overview

The HATT prototype used was a 3-bladed, direct-drive turbine. The rotor blades are 0.385 m in length and were designed using a Wortmann FX 63-137 aerofoil profile. A schematic of the rotor and blade profile can be seen in Figure 2. The pitch angle was set at 8° for all tests. The blade root design was selected to provide a smooth transition to the hub, adequate structural support, and to house the blade root strain gauges. Full details of the turbine design and hub geometry are given in [18,19]. The turbine was equipped with a Bosch Rexroth permanent magnet synchronous machine (PMSM) which was configured to operate in either constant speed or torque mode. Either speed or torque control strategies were applied for all the tests discussed in this paper.

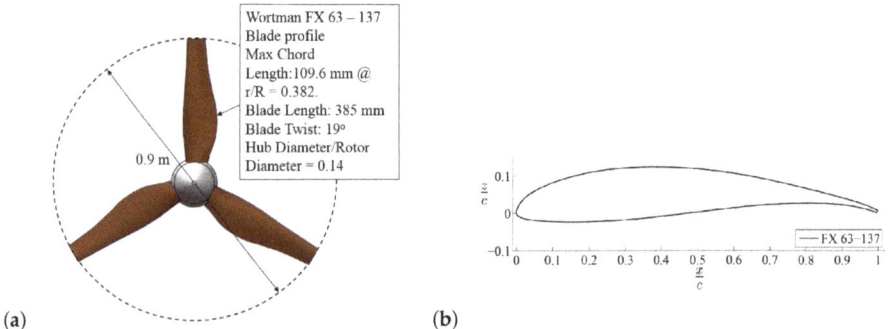

Figure 2. Rotor schematic for the 0.9 m HATT used for the testing (**a**) and Wortman FX 63-137 profile used to create the turbine blade geometry (**b**).

To monitor power capture, the induced rotor torque was quantified from the torque generating current (TGC) measured at the motor. The TGC is the quadrature axis current (iq) required by the motor to generate a specified torsional load (τ_{mot}). The relationship between the TGC and the torque developed by the motor is related by the, 'motor torque constant', which in the case of the PMSM used was specified by the manufacturer as 6.60 N m A^{-1}. Under steady state operation, the motor torque is equal and opposite to the torque developed by the rotor blades (τ_{rotor}) when operating in the tow tank.

This can be seen in the 'swing' equation for the scale model HATT drive train which is presented in Equation (1).

$$J\ddot{\theta} = \tau_{rotor} - D(\dot{\theta}) - \tau_{mot} \tag{1}$$

where, J is the moment of inertia of the turbine drive shaft and rotor in kg m^2 and θ is the rotational position of the turbine rotor. τ_{rotor} is the mechanical torque corresponding to the rotor and τ_{mot} is the motor torque. In the preceding discussion steady state operation refers to the operation whereby the rotational acceleration of the rotor is approximately zero, i.e., $\ddot{\theta} \simeq 0$. Under set-point speed control operation, if the turbine rotor produces a large torque during operation, the motor will have to apply a torque in the opposite direction to maintain the specified rotational velocity. Conversely, if the turbine rotor produces a small positive torque then the motor torque will be reduced, again to maintain the zero rotational acceleration objective inherent in speed control operation. Dynamic friction and damping in the drivetrain due to seals and bearing losses was quantified to negate the effects of the lumped dynamic friction and damping factor D in Equation (1). The dynamic friction and damping was obtained by measuring the TGC at several different rotational speeds (rpm) in still water without the blades attached to the hub. The rotor torque was computed as:

$$\tau_{rotor} = ((-5.3 \times 10^{-6} \times \text{rpm}^2) + (0.0015701 \times \text{rpm}) - 0.1043073) - (\text{TGC} \times 6.6) \tag{2}$$

The turbine housed an optical encoder for position and rotational velocity measurements. The through-bore encoder was mounted on the drive shaft in close proximity to the PMSM. The encoder is a Heidendain ECN 113. The accuracy of the instrument specified by the manufacturers is $\pm 0.0056°$. To monitor the thrust loading on the turbine rotor, an Applied Measurements thrust transducer was fitted on the driveshaft downstream of the rotor hub but upstream of seals and thrust bearings. Calibrations were provided for each direction by the manufacturer. The offset in the calibrations was checked prior to each test run from the recordings made with the turbine and tow carriage stationary. This constant was then taken as the actual offset for the calibration for each test, along with the gradient from the supplied calibrations. Noise in the raw thrust signal was removed prior to analysis by using a consistent gradient filter and maximum and minimum cut off values.

Each blade root was also strain gauged to measure the thrust-wise bending moment applied to the blade. Due to a malfunction of the measuring system in blade 1 only the measurements from the roots of blades 2 and 3 are discussed for the tow-only and regular wave cases.

Data acquisition from all channels (turbine, carriage, flow measurement equipment) were synchronized in terms of their start time using a trigger. Data capture was set at 100 Hz for the flow measurement, carriage, rotor thrust, and blade strain gauge measurements. Data from the motor itself (TGC, rotor speed and rotor position) was sampled at 50 Hz.

2.2. Control Strategy

As discussed, a PMSM was used to provide braking torque (in the torque control case) and to maintain a set-point turbine rotational velocity (in the speed control case). The PMSM was managed by a drive section. The three-phase supply (420 V, 50 Hz) was rectified from AC to DC through a Voltage Source Converter (VSC). An inverter converted the DC bus voltage to a three-phase AC voltage which was connected to the motor. Both the rectifier and inverter were connected via a DC bus integrated with a DC bus capacitor, the voltage at the DC bus was maintained at 750 V. The power flow to and from the motor was managed by the VSCs either side of the DC bus—similar to back-to-back setup used for tidal stream and wind turbines adopting a direct-drive PMSM topology. The VSC operated with a switching frequency of 4 kHz. The use of back to-back VSCs and the encoder permitted the use of a servo-based vector-oriented control (VOC) strategy to regulate the turbine's torque. An additional velocity control loop was also included to set the desired rotational velocity; thus, providing the ability to select the appropriate control strategy; i.e., speed or torque, during the testing campaign. Both control strategies used close-loop control. In the case of the torque control setup the current in the motor phases was measured via a hall effect probe and feedback into the control system to generate a phase current error which is feed through a PID controller. Likewise, in the speed control case the turbine rotational velocity was measured via the encoder and used as feedback into the velocity control loop. Again, the error signal was corrected via a PID controller.

The set-point rotational velocities and applied braking torsion maintained by the motor for each test were selected to undertake testing at a range of tip speed ratio settings (as defined in Section 2.5). The PID controller in both the speed and torque control loops were tuned with guidance from the motor and PLC supplier to achieve critical response by setting appropriate proportional gain values and integral action times.

2.3. Test Program and Procedures

The test program consisted of tow-only, two regular wave-tow, and irregular-wave-tow cases. The tow speed was set equal to 1 m/s for all the cases, and was designed to provide approximately Reynolds independent conditions, $RE_{chord} = 7.9 \times 10^4$ (based on chord length at a radial distance of 70% from root to tip). The regular wave-tow cases were divided in two waves types; waves that provided extreme loading conditions to the turbine prototype and here they are referred to as 'extreme wave-tow' and characteristic regular waves that corresponded to the significant wave height (HS) and peak wave period (TP) of the irregular wave cases studied here. The second type of regular wave-tow are simply referred to here as 'regular wave-tow'. The irregular wave cases are referred to as 'Irregular wave-tow'. The wave height (H) of the regular wave-tow cases was limited by the highest peaks in the irregular wave case not exceeding the maximum possible wave height in the facility (0.45 m). A Jonswap spectrum was used for the irregular waves as a starting point to simulating more realistic wave conditions, with relevance to projects in the North Sea. Table 2 shows a summary of the given test parameters.

Table 2. Test matrix showing the wave cases and control type used throughout the test campaign. All tests were run at 1.0 m/s.

Wave Case	Wave Height (m)	Wave Period (s)	Control Type
Tow-only * Speed Control Tow	N/A	N/A	Speed
Tow-only * Torque Control Tow	N/A	N/A	Torque
Regular wave-tow * Speed Control Wrg	0.19	1.44	Speed
Regular wave-tow * Torque Control Wrg	0.19	1.44	Torque
Irregular wave-tow * Speed Control Wjsp	0.19	1.44	Speed
Irregular wave-tow * Torque Control Wjsp	0.19	1.44	Torque
Extreme wave-tow * Speed Control Wex	0.40	2.00	Speed
Extreme wave-tow * Torque Control Wex	0.40	2.00	Torque

* Data label used in figures.

For each flow condition, tests were run in speed control mode and then in torque control mode, with the set speed and torque values varied between test runs to simulate a range of tip speed ratios for each flow condition. Repeat tests were conducted where time allowed to check for consistency in the measurement systems and estimate uncertainty in the results. Table 3 shows the cases that were repeated throughout the testing campaign for a specific flow condition and control strategy.

Table 3. Number of tests done for a specific case.

Speed/Torque Control	Tow	Wrg	Wex	Wjsp
30 rpm			2	
67 rpm		2		
76 rpm	2	2	2	2
5 Nm	2			
10 Nm	2			
15 Nm	2	2		
17.6 Nm	2			
18 Nm	3			
18.8 Nm	2			

Duration of the Tests

Prior to each test, a recording was taken with stationary blades and carriage, to check initial readings on the instrumentation. To start a test the desired speed or torque set-point was selected and then the carriage was started.

The durations of the tow-only tests were approximately 100 s. The repeated tests were recorded for 50 s due to time constrains. The regular wave tests were initiated with the carriage positioned halfway along the tank, compromising on test length but maximizing the time before reflections reached the carriage. This resulted in approximately 30 wave forms or 40 s of useable data.

For the irregular waves, the test time was maximized to better replicate the Jonswap spectrum, so full carriage runs were completed without mitigating reflections from the beach. Thus, it was possible to capture nearly 120 s of data for the irregular wave cases. While not ideal, this was deemed appropriate for preliminary testing in irregular waves to gain insight and start to build knowledge to feed into further testing campaigns during the project where longer test times and reduced reflections will be possible.

2.4. Flow and Wave Measurements

The surface elevation during wave tests was measured primarily by using a capacitance type wave gauge placed in-line with the turbine hub in the cross-stream direction. The distance between the turbine hub and the wave probe was approximately 1.0 m. This was calibrated by setting the gauge to a series of known positions in still water. The average wave height recorded across all regular wave-tow tests was 0.19 m, with a standard deviation of 0.001 m between tests. The average wave period was 1.44 s, with a standard deviation of 0.001 s between tests. The average wave height recorded across all extreme wave-tow tests was 0.39 m, with a standard deviation of 0.002 m between tests. The average wave period was 2 s, with a standard deviation of 0.002 s between tests. A sample of the wave forms obtained for the regular and extreme waves obtained in two tests can be observed in Figures 3 and 4.

It can be noticeable in Figures 3a and 4a that the wave heights obtained for each of the regular and extreme tests were consistent. The average wave height across the entire length of the signal for each of the regular wave-tow cases had a deviation of 2.8% of the mean value whereas a slightly higher variability was observed for the extreme wave-tow cases with a variation of nearly 4% of the mean value.

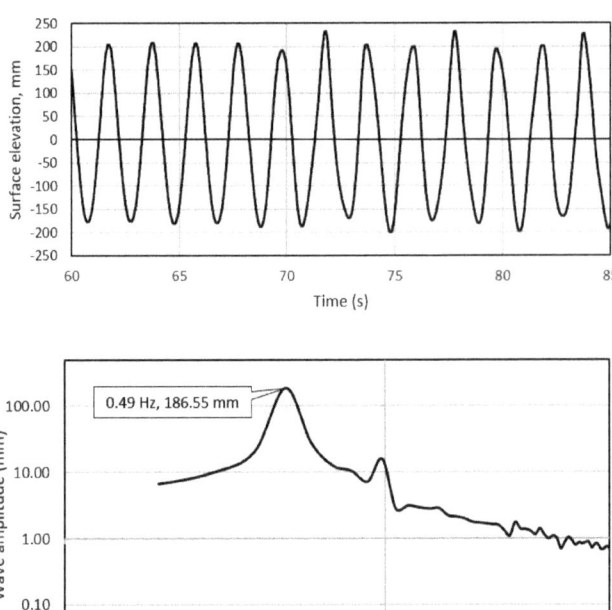

Figure 3. Surface elevation of one of the Extreme wave-tow cases, demonstrating an average wave height of 0.39 m (**a**,**b**) the frequency spectrum of the signal.

For the irregular wave tests, the records were analyzed in the time domain and the zero up-crossing points were selected to define each wave period. Reasonable agreement between the average wave period and wave height was obtained between repeated test runs. However, variations between the repeated records in terms of the maximum and minimum wave periods were noticeable. The average wave period and wave height are also smaller than the programmed significant wave height and peak period, as it can be seen in Table 4.

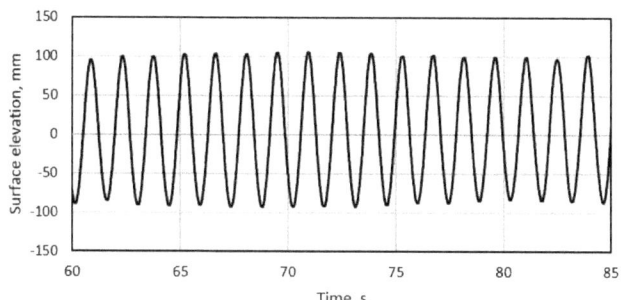

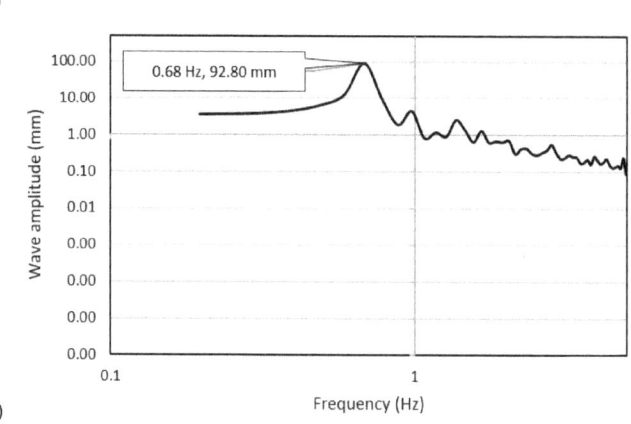

Figure 4. Surface elevation of one of the Regular wave-tow cases, demonstrating an average wave height of 0.19 m (**a**,**b**) the frequency spectrum of the signal.

Table 4. Characteristics of the irregular wave tests using the zero up -crossing method.

Parameter	Speed Control	Speed Control (Repeat)	Torque Control	Torque Control (Repeat)
Length of data (s)	136	140	135	139
Average wave period (s)	1.25	1.24	1.25	1.29
Maximum wave period (s)	1.83	2.25	1.96	3.14
Minimum wave period (s)	0.25	0.45	0.08	0.58
Average wave height (m)	0.12	0.13	0.12	0.13
Maximum wave height (m)	0.28	0.33	0.30	0.29
Minimum wave height (m)	0.00	0.01	0.00	0.01

While the irregular wave forms in this study are somewhat arbitrary in nature, these do still allow the effect of dynamic changes in wave height and period on the turbine loading to be investigated. They will provide a preliminary insight into the possible effects of random wave patterns on the turbine in contrast with regular waveforms. A picture of the surface elevation recorded during the irregular wave tests can be observed in Figure 5.

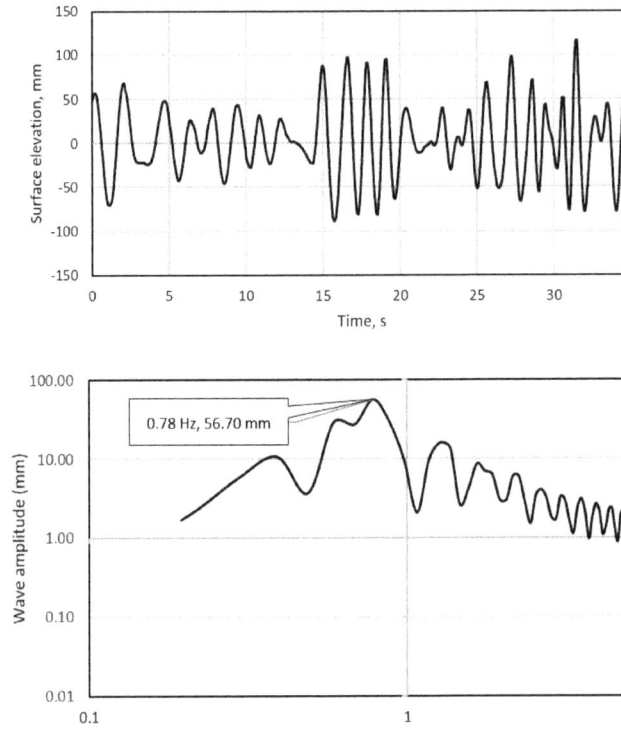

Figure 5. Water surface elevation time series for the irregular wave case undertaken (**a**,**b**) the frequency spectrum of the signal.

The carriage velocity reading will be used to provide an average speed, which should suffice considering that wave-current interaction effects should not be present in a tow tank, resulting in zero mean effect on the 'current'. When investigating the data in the time domain the wave height measurements will be used to represent the flow conditions.

2.5. Data Processing

The analysis of the data will be described in terms of the following non-dimensional parameters:

$$C_P = \frac{P}{0.5 \rho A V^3} \quad (3)$$

$$C_T = \frac{T}{0.5 \rho A V^2} \quad (4)$$

$$C_M = \frac{M}{0.5 \rho A r V^2} \quad (5)$$

$$TSR = \frac{\omega r}{V} \quad (6)$$

where P and T are the average hydrodynamic power and thrust generated by turbine. The power is calculated using the TGC, as previously explained in Section 2.1 multiplied by the turbine angular velocity ω in rad/s; thus, given a final output in Watts. M is the blade root bending moment. The swept area of the rotor including the central hub is approximately 0.64 m^2, deriving it from the turbine radius (r) of 0.45 m. The density of the water was considered in these calculations as 999 kg/m^2. Both power (C_P) and thrust (C_T) coefficients are related in the results section to the tip speed ratio (TSR). This non-dimensional value defines the ratio between the blade tip speed ($\omega \times r$) and the tow velocity (V), as shown in Equation (6). The flow velocity used in these experiments was 1.0 m/s.

The loading fluctuations analysis will be carried out by using maximum and minimum values of the time domain signals. The data processing will also include the frequency domain analysis of the signal.

3. Results and Discussion

The results from the turbine thrust, torque, and blade root bending moment measurements are discussed in this section. Section 3.1 describes the results of power and thrust in terms of non-dimensional parameters, as discussed in Section 2.5. Section 3.1 also includes the average values of the blade root bending moments. Section 3.2 presents the results as average fluctuations per wave period. These mean fluctuations were obtained by calculating the average of the discrete signal periods (peak to peak range). The spectral analysis is presented in Section 3.3 to investigate the relation between the amplitude of the peak frequencies and the control strategies.

3.1. Non-Dimensional Power Curves and Blade-Bending Moments

The resultant non-dimensional C_P and C_T performance curves for each of the flow conditions and control strategy described in Table 2 can be seen in Figures 6 and 7. It must be noted that the scatter in both figures is considerably low, especially since the results include repeated tests done for the same operating point. Please note that there is excellent agreement between the repeated values. The standard deviation between three repeated tests was less than 1% of the mean value for each of the power and thrust coefficients for the 18 Nm (TSR $\sim$ 3) torque control test (tow-only condition). Similarly, two tests were done for a speed control mode at 76 rpm (TSR $\sim$ 3.6) operating under tow-only conditions, the resultant standard deviation between tests was 1.2% of the mean value for the thrust coefficient and less than 1% for the power coefficient. For the extreme and regular wave-tow tests, the variability between repeated tests was also lower than 1% of the mean value.

The dispersion of the average values between tow-only and wave cases for the same operating point was in the order of 1.3% of the mean value for the thrust coefficient and power coefficient (for a TSR $\sim$ 3.6). These values also considered torque and speed control cases, demonstrating that mean values were not affected by the control strategy nor the influence of either the regular or irregular wave cases.

In Figure 8 the mean values of the root bending moment coefficients of blade 2 and blade 3 are given for each of the cases. The blade 2 root bending moment coefficients are about 6% higher than those obtained from the output of blade 3; however, it is clear that they present a similar trend. As the values of blade 2 are constantly greater independently of the rotational speed or of the turbine control mode, it is likely that this increase in bending moment is only due to a slight disparity between the instrumentation and waterproofing methods used for each connection.

Additional information regarding the repeatability of the tests can be related to the dispersion of the rotational speeds achieved when implementing the speed control mode or the variation of the TGC, when referring to the torque control strategy. The standard deviation in the rotor rotational speed values during speed control mode was found to be less than 0.25 rpm. Similar to the dispersion of C_P and C_T measurements, this is about 1% or less of the mean rotational speed. The standard deviation for the TGC was about 3% of the average values of the torque control mode for all the cases. The calculations for both the rotational speeds and the TGC dispersion values were not altered by the incorporation of waves.

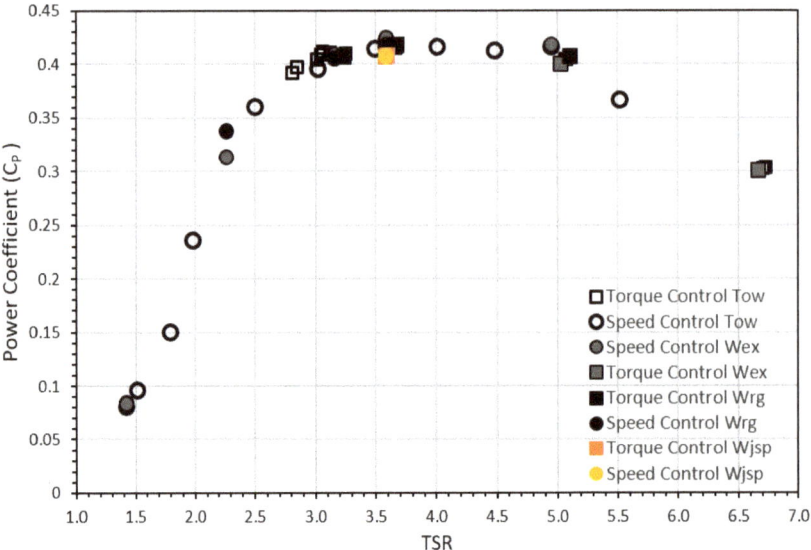

Figure 6. Power Coefficient (C_P)—Tip speed ratio (TSR) curve including all the test cases for both control strategies.

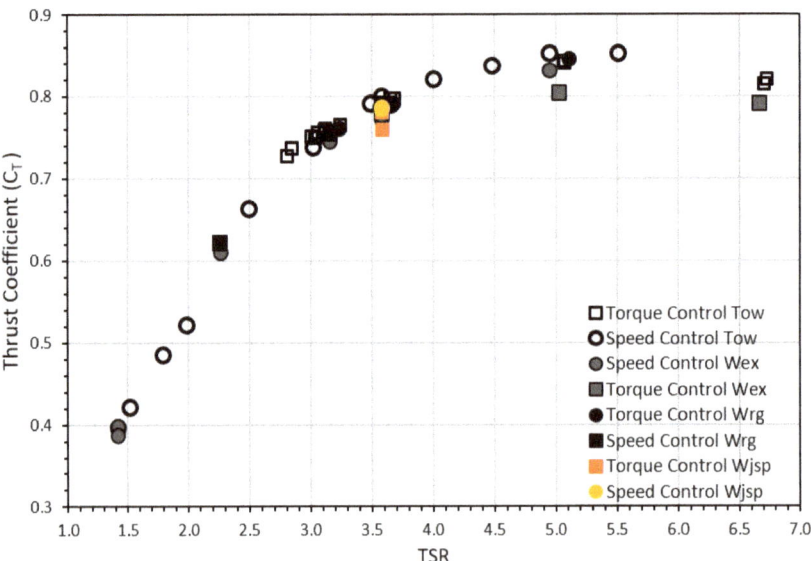

Figure 7. Thrust coefficient (C_T)—Tip speed ratio (TSR) curve including all the test cases for both control strategies.

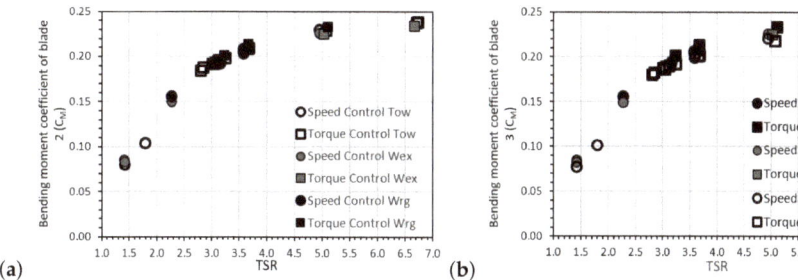

Figure 8. Average blade root bending moment coefficient (C_M) for blade 2 (**a**) and blade 3 (**b**) versus tip speed ratio (TSR). Please note that there is not available for the irregular wave-tow tests.

It is noticeable that at a TSR ~ 5, a relatively large deviation in the mean C_T values for the tow-only and regular wave cases relative to the extreme wave case can be seen when the torque control operation is used. The authors believe this discrepancy is likely to be resultant from the large oscillations in TSR value observed under this test case. The large oscillations in rotational velocity and fluid velocity (due to the wave motion and control strategy) mean a wide range of TSR can be observed for this case. Due to the inherent shape of the C_T curve, which exhibits a relatively small gradient at the maxima region (4.5 < TSR < 5.5) and a relatively steep increase at lower TSR values, this results in a positively skewed data sets leading to a slightly lower mean thrust value.

The main objective of this investigation was to compare the repercussions of using two control strategies in the loading of a HATT. Therefore, an equivalent parameter to reach the same operational point of the turbine was required to be met. For example, to reach a TSR of 3.5 a constant speed of 76 rpm must be met when operating in a speed control mode or a torque of 15 Nm when using a torque control mode. Therefore, when compared torque and rotational speed for a similar case using two control strategies, in this instance 76 rpm and 15 Nm tests, a variation of around 1.9 rpm and 2 Nm between them was achieved.

3.2. Time Average Signal Fluctuations

Figure 9a shows the average turbine rotational velocity fluctuation per wave period plotted against TSR for the torque control cases. The average maximum and minimum values of rotational velocity per wave period compared to the TSR are also included in Figure 9b for the same set of cases. The average maxima and minima observed are highlighted by the upper and lower error bars, respectively. The speed control cases have been omitted in Figure 9 as the oscillations in rotational velocity are inherently limited by the control strategy adopted.

It is clear in Figure 9 that the size of the fluctuations relative to the mean rotational velocity decreases with average rotor speed, but both maximum and minimum speed increase with increased average speed. The fluctuation range is about 57% between the mean value of the rotational speed and the rotational speed range when looking at the regular wave-tow cases. The fluctuations in the two repeated irregular wave cases is ~34% of the mean speed. A far bigger fluctuation range relative to the mean rotational velocity is observed in the extreme wave-tow case (~110%) when compared with the regular wave-tow case (~35%) for the comparable TSR value of 5.1.

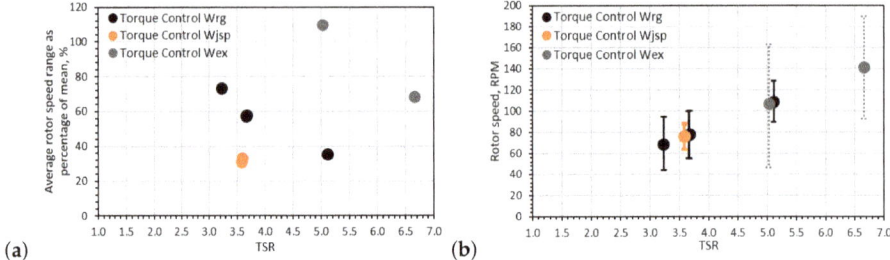

Figure 9. Average turbine rotational velocity fluctuation per wave period only for torque control cases: (**a**) the rotational velocity fluctuation range as a percentage of the mean rotational velocity for each case is presented and (**b**) the average maximum and minimum turbine rotational velocity per wave period are indicated by the extremes of the error bars.

In Figure 10 the fluctuations in the torque signals are presented, this time only for the speed-controlled cases. There is a modest increase in the torque fluctuation range with rotational velocity (Figure 10a) with a maximum fluctuation range of about 45% of the mean torque value in regular wave-tow tests. This increase in torque range is more evident for the extreme wave-tow cases with a fluctuating values of between 113% and 120% of the mean torque compared with a 45% fluctuation in the regular-tow case and a 35% fluctuation in the irregular-tow case, again for the comparable TSR of ∼3.6. It should be noted, at the aforementioned TSR, that the extent of fluctuations is relatively scattered for repeated extreme and irregular wave-tow cases, despite repeatability in the mean values observed. The maximum fluctuation in rotor torque was observed at high thrust conditions under the extreme-tow case with fluctuations of ∼160% of the mean value at TSR ∼ 4.9.

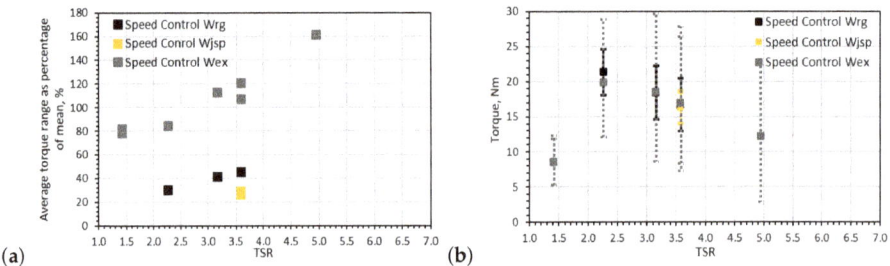

Figure 10. Average rotor torque fluctuation per wave period only for speed control cases: (**a**) the torque fluctuation range as a percentage of the mean rotor torque for each case is shown and (**b**) the average maximum and minimum rotor torque per wave period are indicated by the extremes of the error bars.

The average rotor thrust fluctuations per wave period are plotted in Figures 11 and 12 for the torque and speed control cases, respectively. It is clearly visible that using the torque control mode induces higher thrust fluctuations in the three wave-tow cases. It is visible that there is an opposite trend in the fluctuations occurring in the rotor thrust when using a specific control type. When the speed control mode was used, a linear increase in the variation of the thrust loading was observed in both wave-tow tests whereas a linear decrease in thrust fluctuations is observed when the torque control mode is applied. This contradicts the information reported by [5], who employed a torque control strategy in the experimental campaign and found an increase in the fluctuating thrust coefficient with TSR. However, it is somewhat difficult to compare both results as the fluctuating thrust coefficient shown in [5] was only indicated by a standard deviation value rather than a difference in amplitude between peak values, as it is presented here. Moreover, there are other differences between studies such as waveforms, flow velocities, facility types, and rotor types which will contribute to discrepancies between studies.

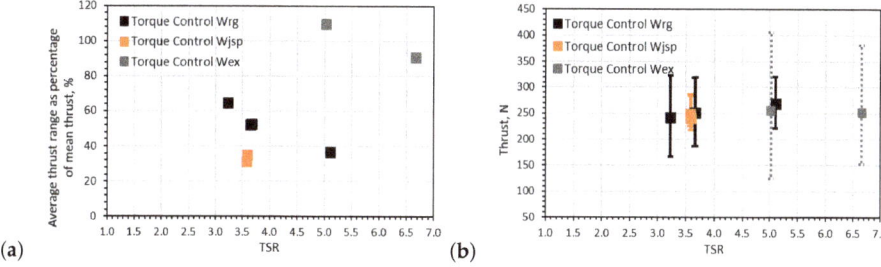

Figure 11. Average rotor thrust fluctuations per wave period only for torque control cases: (**a**) the thrust fluctuation range as a percentage of the mean rotor thrust for each case is presented and (**b**) the average maximum and minimum rotor thrust per wave period are indicated by the extremes of the error bars.

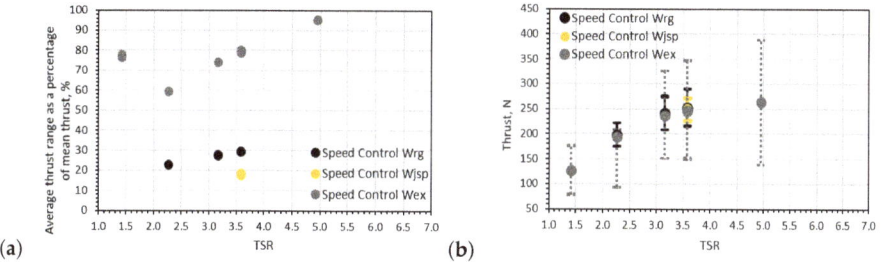

Figure 12. Average rotor thrust fluctuations per wave period only for speed control cases: (**a**) the thrust fluctuation range as a percentage of the mean rotor thrust for each case is presented and (**b**) the average maximum and minimum rotor thrust per wave period are indicated by the extremes of the error bars.

For a similar operating point (TSR $\sim$ 3.15), the thrust fluctuation of the rotor increased by 57% when applying torque control ($\sim$65% fluctuation relative to the mean value) instead of speed control ($\sim$30% fluctuation relative to the mean value) for the regular wave-tow case. Comparing the effect of control type for the extreme wave-tow case at TSR $\sim$ 5.0, it can be seen that again an increase in thrust fluctuation was observed when using torque control; that is a 115% average fluctuation observed in the torque control case relative to a 95% fluctuation in the speed control case.

The results would suggest a greater sensitivity of thrust load variations to control strategy at lower TSRs. This phenomenon would appear to arise from the roughly asymptotic nature of the thrust coefficient curve which is specific to the rotor setup tested; as such, the result may be somewhat rotor specific which would also explain differences when comparing the results with other studies. Evidence for this reasoning can be seen in Figure 11b where the maximum thrust per wave period observed for differing TSRs for the regular wave-tow case were approximately constant for the three cases tested. While this reasoning maybe rotor geometry specific, this finding leads to the possible notion of developing rotor designs to mitigate against thrust fluctuations at peak power. Further tests to compare thrust fluctuations in the peak power region for the extreme wave-tow case and the differing control types were not feasible in these experiments due to the turbine stalling at TSRs lower than 5.0. The rotor stalling in the torque control case was due to operating momentarily at TSRs lower than the peak torque TSR for the rotor causing the turbine to stall.

Lastly, there appears to be a correlation between wave height and thrust range, with fluctuation size increasing with average wave height, as the average wave height in the irregular wave time series was lower than that of the regular wave-tow case. However, more experiments are required to fully understand the loading patterns associated with control strategies in random waves.

Figures 13 and 14 show the average bending root moment fluctuations per wave period only for torque and speed control cases, respectively. The trend between the average values of the bending moment fluctuations of blade 2 and blade 3 is almost identical. Hence the reason of only using the values of blade 2 to show the maximum and minimum values on Figures 13b and 14b. The pattern of the fluctuations when applying torque or speed control is similar to that presented in Figures 11 and 12 with fluctuation size increasing with TSR in speed control mode but decreasing with TSR in torque control mode.

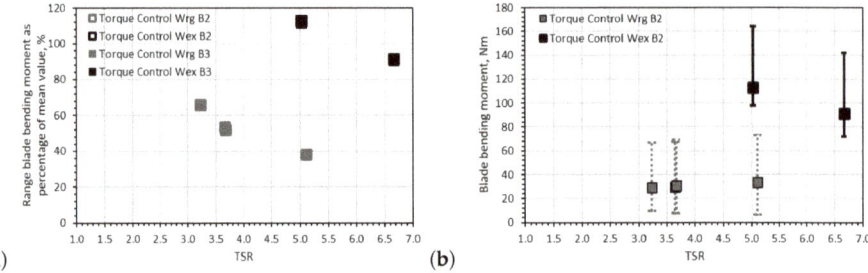

Figure 13. Average bending root moment fluctuations per wave period only for torque control cases: (a) the bending root moment fluctuation range as a percentage of the mean bending moment for each blade is presented and (b) the average maximum and minimum rotor thrust per wave period are indicated by the extremes of the error bars (only for blade 2).

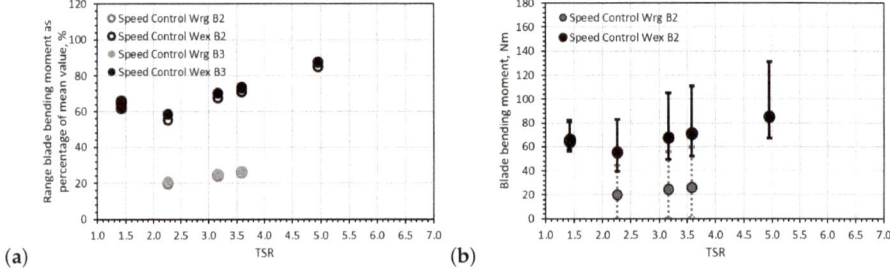

Figure 14. Average bending root moment fluctuations per wave period only for speed control cases: (a) the bending root moment fluctuation range as a percentage of the mean bending moment for each blade is presented and (b) the average maximum and minimum rotor thrust per wave period are indicated by the extremes of the error bars (only for blade 2).

When comparing the corresponding fluctuations for the same operating region, it was found that the bending moment fluctuation of the blade increased from when applying torque control the torque control mode increases the average bending moment to 53.5% when compared to that obtained in the speed control experiment (26.5%) at TSR = 3.6, in regular wave-tow tests. A small increase of 26% was observed for the only case available for the extreme wave-tow experiment between the torque and speed control. This percentage is substantially lower than that presented by the regular wave-tow cases; however, this was for an operating point of TSR = 5.1 ($\simeq$105 rpm). It is interesting that the difference in loading fluctuations between torque and control tests for TSR = 5 are substantially lower than those associated with the regular wave-tow cases, especially since it was depicted in the C_P—TSR curve that the peak power can be achieved between 3.0 < TSR < 5.0. It is clear that additional tests related to torque and control strategies must be performed when a turbine operates under wave and current conditions within this region.

3.3. Frequency Domain Analysis

Figures 15 and 16 show the spectral analyses of the torque, power, thrust and root bending moments for toque and speed control conditions for a regular wave-tow case at TSR = 3.6. The dominant frequency observed for most of the plots corresponds to the wave frequency of 0.69 Hz (wave period of 1.44 s). When using speed control, the presence of the wave frequency in the spectrum of the torque signal is evident but unnoticeable when the torque control mode was on. Therefore, it was deemed necessary to include a plot to relate the influence of both torque and speed control strategies in power. It can thus be observed in Figure 15b, that with constant torque the amplitude of the power at the wave frequency is higher. This means that in a scaled-up device, the fluctuating power on the grid will be higher when a turbine is subjected to wave conditions. A similar trend is observed for the spectra of the thrust and blade root bending in Figure 16. However, it is interesting to note that for the extreme wave cases, the power amplitudes are substantially lower than those obtained at constant speed conditions, see Figure 17a. Unfortunately, the experiments related to extreme wave-tow tests were limited, due to the capabilities of the turbine at such extreme conditions. However, in the future it will be contemplated to repeat a similar but more conservative wave-tow condition to confirm the outcomes of this investigation. The results are of relevance as the this means that at this operating point (TSR = 5, C_P = 0.4 and C_T = 0.8) on a larger scale, the drivetrain and rotor will be subjected to smoother mechanical conditions when operating in harsher environments, compared to those achieved at peak conditions (TSR = 3.6, C_P = 0.42 and C_T = 0.78).

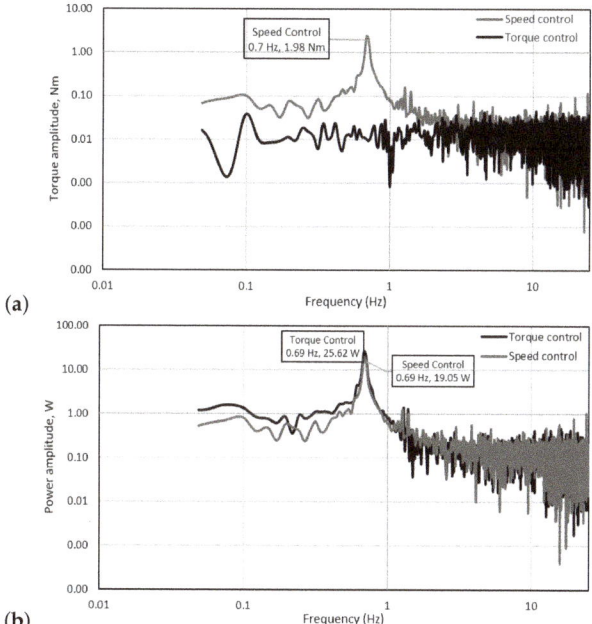

Figure 15. Frequency domain graphs for a regular wave-tow test at peak power condition of TSR = 3.6, i.e., 76 rpm, 15 Nm for rotor: (**a**) torque and (**b**) power.

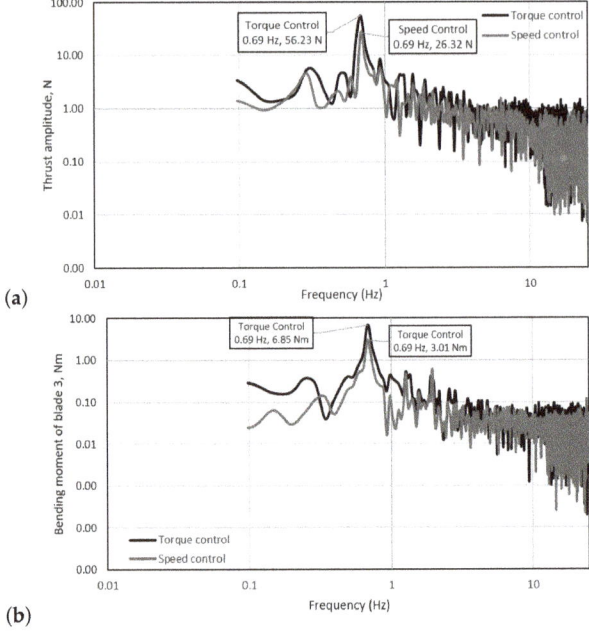

Figure 16. Frequency domain graphs for a regular wave-tow test at peak power condition of TSR = 3.6, i.e., 76 rpm, 15 Nm for: (**a**) thrust and (**b**) root bending moment blade 3.

It is clear that the peak frequencies developed during the torque control tests are higher than those seen in the speed control tests, at least for the regular wave-tow cases shown in Figures 15 and 16. On a smaller scale, the frequencies of the rotational speed are also visible in the power and torque spectra, but the amplitudes are substantially low compared to the magnitude of the wave frequency. The same is depicted in the thrust and bending moment spectra with the addition of the frequency related to the rotational speed in combination with the wave frequency ($\simeq$1.96 Hz). Again, the amplitudes of the additional frequencies are considerably lower than those related to the wave frequency. This work is however purely focused on the interaction between in-line waves and currents which may not be fully representative of a tidal site, as demonstrated by [20]; where it was found that between 49–93% of the time the waves propagate with a direction of 20 degrees of tolerance. Therefore, the directionality of waves and currents is of importance and should be incorporated in future experiments to understand the loading on tidal turbines subjected to these conditions.

In Figure 17 the peak amplitudes of corresponding wave frequencies are shown for power, thrust, and root bending moment of blade 2 and 3. For the thrust and root bending moments of blades 2 and 3, the pattern observed is similar to that presented in Figures 13 and 14 where the fluctuation amplitude increases with rotor velocity in speed control mode, but decreases with rotor velocity in torque control mode.

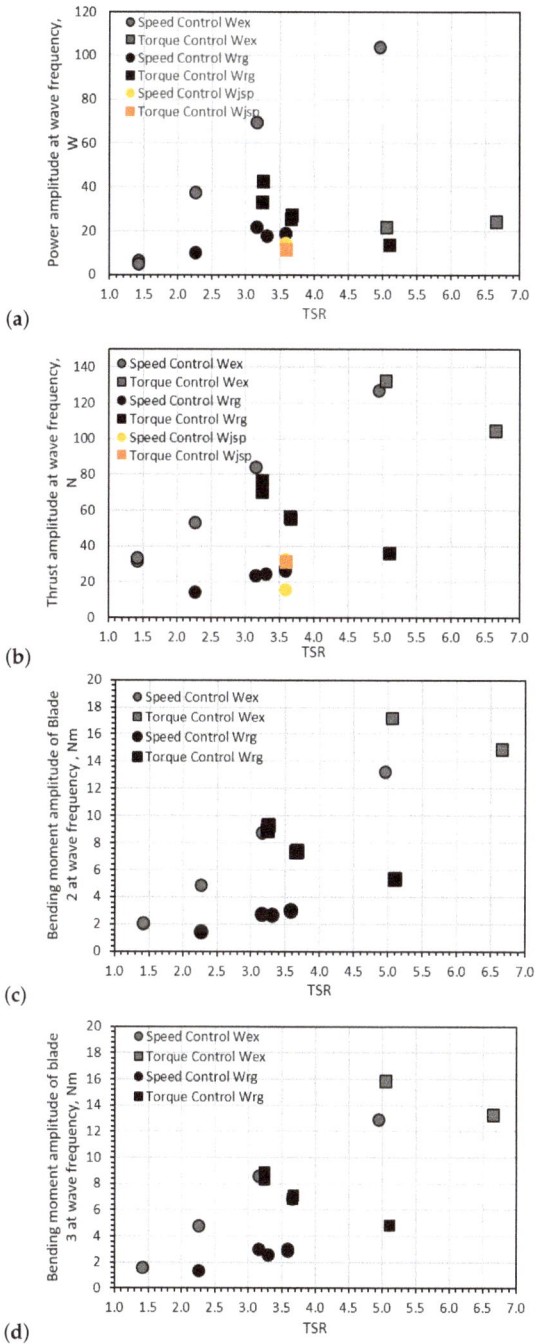

Figure 17. Amplitudes observed at wave frequencies of: (**a**) power, (**b**) rotor thrust, (**c**) root bending moment blade 2 and (**d**) root bending moment blade 3.

4. Conclusions

The influence that control strategies have on the loading of a three-bladed horizontal-axis HATT was investigated in a wave-tow tank. The turbine was tested under two types of regular wave-tow cases and an irregular wave-tow test. It was found that when looking at average values, the control mode or the type of wave-tow case investigated did not affect the rotor thrust, rotor torque or blade root bending moment measurements when these where compared to tow-only conditions. The thrust and torque fluctuations were substantial under the regular, extreme, and irregular waves cases, with the average peak loads adding up to 30% to the mean value under the conditions tested. However, a greater increase to the mean of nearly 100% was observed when looking at thrust and torque fluctuations developed when the turbine was tested under extreme wave-tow conditions, showing the relation between loading and wave height. For the test cases where the set operational point was reached in both torque and control strategies, it was found that the thrust and root blade-bending moment fluctuations per wave period were in the order of 40% higher when the torque control was applied compare to the thrust control tests. A pattern showing that the thrust and blade root bending moments was observed in both time and frequency domain where the fluctuation or peak amplitude of the thrust or bending moment increased with rotor velocity in speed control mode but decreased with rotor velocity in torque control mode. The power spectra were also studied. It was observed that for the extreme wave-tow cases the amplitude of the wave frequency was substantially lower when using torque control than constant speed. The study shown here highlights the importance for understanding power control, smoothing and integration with the grid, and structural and fatigue design of components such as the blades, driveshaft, generator etc. There is certainly more need to investigate the matter, especially how these effects follow a similar pattern in an environment with wave directionality and turbulence. It is also important to recognize that the experiments undertaken for the irregular wave cases were quite limited. This matter will also be studied in the future in laboratory conditions where longer time series are achievable. Given the fact that the amplitude fluctuations in the power spectra were exceptionally low when applying torque control, further experiments will also be contemplated in the future.

Author Contributions: Conceptualization, M.A. and S.O.-S.; methodology, K.P. and M.A.; software, M.A.; validation, S.O.-S., M.A. and K.P.; formal analysis, S.O.-S., M.A. and K.P.; investigation, I.S.; resources, C.L.; data curation, R.E.; writing–original draft preparation, S.O.-S.; writing–review and editing, S.O.-S. and M.A.; supervision, T.O. and C.J.; project administration, T.O. and C.J.; funding acquisition, M.A., T.O. and C.J.

Funding: This research was funded by EPSRC grant number EP/N020782/1. Facility access was funded by the H2020 MARINET 2 program.

Acknowledgments: This investigation was funded The authors' would like to thank the staff at CNR-INM for their expertise and support during the testing period.

Conflicts of Interest: The authors declare no conflict of interest.

Abbreviations

The following abbreviations are used in this manuscript:

BEMT	Blade Element Momentum Theory
CFD	Computational fluid dynamics techniques
H	Wave height
HATT	Horizontal Axis Tidal Turbine
HS	Significant wave height
iq	Quadrature axis current
PID	Proportional–integral–derivative
PMSM	permanent magnet synchronous machine
rpm	Revolutions per minute
TGC	Torque generating current
TP	Peak wave period

TSR   Tip Speed Ratio
VOC   Vector-oriented control
VSC   Voltage Source Converter

References

1. Nevalainen, T.M.; Johnstone, C.M.; Grant, A.D. A sensitivity analysis on tidal stream turbine loads caused by operational, geometric design and inflow parameters. *Int. J. Mar. Energy* **2016**, *16*, 51–64. [CrossRef]
2. Nevalainen, T.; Peter, D.; Johnstone, C. Internal bearing stresses of horizontal axis tidal stream turbines operating in unsteady seas. In Proceedings of the 3rd Asian Wave and Tidal Energy Conference, Singapore, 24–28 October 2016.
3. Tatum, S.C.; Frost, C.H.; Allmark, M.; O'Doherty, D.M.; Mason-Jones, A.; Prickett, P.W.; Grosvenor, R.I.; Byrne, C.B.; O'Doherty, T. Wave–current interaction effects on tidal stream turbine performance and loading characteristics. *Int. J. Mar. Energy* **2015**, *14*, 161–179. [CrossRef]
4. Holst, M.A.; Dahlhaug, O.G.; Faudot, C. CFD Analysis of Wave-Induced Loads on Tidal Turbine Blades. *IEEE J. Ocean. Eng.* **2015**, *40*, 506–521. [CrossRef]
5. de Jesus Henriques, T.A.; Tedds, S.C.; Botsari, A.; Najafian, G.; Hedges, T.S.; Sutcliffe, C.J.; Owen, I.; Poole, R.J. The effects of wave–current interaction on the performance of a model horizontal axis tidal turbine. *Int. J. Mar. Energy* **2014**, *8*, 17–35. [CrossRef]
6. Gaurier, B.; Davies, P.; Deuff, A.; Germain, G. Flume tank characterization of marine current turbine blade behaviour under current and wave loading. *Renew. Energy* **2013**, *59*, 1–12. [CrossRef]
7. Galloway, P.; Myers, L.; Bahaj, A.A. Quantifying wave and yaw effects on a scale tidal stream turbine. *Renew. Energy* **2014**, *63*, 297–307. [CrossRef]
8. Barltrop, N.; Varyani, K.S.; Grant, A.; Clelland, D.; Pham, X. Investigation into wave–current interactions in marine current turbines. *Proc. Inst. Mech. Eng. Part A J. Power Energy* **2007**, *221*, 233–242. [CrossRef]
9. Consul, C.A.; Willden, R.H.J.; McIntosh, S.C. Blockage effects on the hydrodynamic performance of a marine cross-flow turbine. *Philos. Trans. R. Soc. A* **2013**. [CrossRef] [PubMed]
10. Ordonez-Sanchez, S.; Ellis, R.; Porter, K.E.; Allmark, M.; O'Doherty, T.; Mason-Jones, A.; Johnstone, C. Numerical Models to Predict the Performance of Tidal Stream Turbines Working under Off-Design Conditions. *Ocean Eng.* **2018**, submitted.
11. Martinez, R.; Payne, G.S.; Bruce, T. The effects of oblique waves and currents on the loadings and performance of tidal turbines. *Ocean Eng.* **2018**, *164*, 55–64. [CrossRef]
12. Draycott, S.; Payne, G.; Steynor, J.; Nambiar, A.; Sellar, B.; Venugopal, V. An experimental investigation into non-linear wave loading on horizontal axis tidal turbines. *J. Fluids Struct.* **2019**, *84*, 199–217. [CrossRef]
13. Draycott, S.; Nambiar, A.; Sellar, B.; Davey, T.; Venugopal, V. Assessing extreme loads on a tidal turbine using focused wave groups in energetic currents. *Renew. Energy* **2019**, *135*, 1013–1024. [CrossRef]
14. Cooperman, A.; Martinez, M. Load monitoring for active control of wind turbines. *Renew. Sustain. Energy Rev.* **2015**, *41*, 189–201. [CrossRef]
15. Frost, C.; Benson, I.; Jeffcoate, P.; Elsäßer, B.; Whittaker, T. The Effect of Control Strategy on Tidal Stream Turbine Performance in Laboratory and Field Experiments. *Energies* **2018**, *11*, 1533. [CrossRef]
16. Nambiar, A.; Anderlini, E.; Payne, G.S.; Forehand, D.; Kiprakis, A.; Wallace, R. Reinforcement Learning Based Maximum Power Point Tracking Control of Tidal Turbines. In Proceedings of the 12th European Wave and Tidal Energy Conference, Cork, Ireland, 27 August–1 September 2017.
17. Ordonez-Sanchez, S.; Porter, K.; Frost, C.; Allmark, M.; Johnstone, C.; O'Doherty, T. Effects of Wave-Current Interactions on the Performance of Tidal Stream Turbines. In Proceedings of the 3rd Asian Wave and Tidal Energy Conference, Singapore, 24–28 October 2016.
18. Allmark, M.; Ellis, R.; Porter, K.; O'Doherty, T.; Johnstone, C. The Development and Testing of a Lab-Scale Tidal Stream Turbine for the Study of Dynamic Device Loading. In Proceedings of the 4th Asian Wave and Tidal Energy Conference, Scottish, UK, 9–13 September 2018.

19. Ellis, R.; Allmark, M.; O'Doherty, T.; Ordonez-Sanchez, S.; Mason-Jones, A.; Johannesen, K.; Johnstone, C. Design process for a scale horizontal axis tidal turbine blade. In Proceedings of the 4th Asian Wave and Tidal Energy Conference, Scottish, UK, 9–13 September 2018.
20. Lewis, M.J.; Neill, S.P.; Hashemi, M.R.; Reza, M. Realistic wave conditions and their influence on quantifying the tidal stream energy resource. *Appl. Energy* **2014**, *136*, 495–508. [CrossRef]

© 2019 by the authors. Licensee MDPI, Basel, Switzerland. This article is an open access article distributed under the terms and conditions of the Creative Commons Attribution (CC BY) license (http://creativecommons.org/licenses/by/4.0/).

Article

Applying International Power Quality Standards for Current Harmonic Distortion to Wave Energy Converters and Verified Device Emulators

James Kelly [1,*], Endika Aldaiturriaga [2] and Pablo Ruiz-Minguela [3]

[1] MaREI Centre, Environmental Research Institute, University College Cork, Ringaskiddy P43 C573, Ireland
[2] IDOM Consulting, Engineering, Architecture, 48015 Bilbao, Spain; ealdaiturriaga@idom.com
[3] Tecnalia, Parque Científico y Tecnológico de Bizkaia Astondo Bidea, Edificio 700, E-48160 Bizkaia, Spain; jpablo.ruiz-minguela@tecnalia.com
* Correspondence: james.kelly@ucc.ie

Received: 20 June 2019; Accepted: 20 September 2019; Published: 24 September 2019

Abstract: The push for carbon-free energy sources has helped encourage the development of the ocean renewable energy sector. As ocean renewable energy approaches commercial maturity, the industry must be able to prove it can provide clean electrical power of good quality for consumers. As part of the EU funded Open Sea Operating Experience to Reduce Wave Energy Cost (OPERA) project that is tasked with developing the wave energy sector, the International Electrotechnical Commission (IEC) developed electrical power quality standards for marine energy converters, which were applied to an oscillating water column (OWC). This was done both in the laboratory and in the real world. Precise electrical monitoring equipment was installed in the Mutriku Wave Power Plant in Spain and to an OWC emulator in the Lir National Ocean Test Facility at University College Cork in Ireland to monitor the electrical power of both. The electrical power generated was analysed for harmonic current distortion and the results were compared. The observations from sea trials and laboratory trials demonstrate that laboratory emulators can be used in early stage development to identify the harmonic characteristics of a wave energy converter.

Keywords: renewable energy; ocean energy; wave energy; oscillating water column; power quality; current harmonic distortion; IEC standards; modelling

---

1. Introduction

The Paris Climate Agreement highlighted the global consensus that a low carbon energy strategy needs to be established. The development of clean renewable energy represents a piece of the greater low carbon energy mix, and it has been a growing part of the energy sector this millennium. Wind, solar, and hydropower plants accounted for 90% of the installed renewable generation capacity in the European Union in 2017 [1]. The ocean energy sector, which includes wave and tidal energy, has shown promise for further diversifying the European renewable energy portfolio, but the sector remains in the maturation process and has yet to prove commercial viability. One subsection of the ocean energy sector, wave energy, has been making progress towards commercialisation in recent years [2,3]. This progress includes several large-scale, bottom-fixed shoreline oscillating water columns (OWC), which have been successfully grid-connected [4–6].

A boost to the maturation of the ocean renewable energy field has been the development of the International Electrotechnical Commission (IEC) 62600 standardisation for marine energy converters (MEC), which includes wave, tidal, and other current converters. The European Union Horizon 2020 funded Open Sea Operating Experience to Reduce Wave Energy Cost (OPERA) project is a four-year project with multiple objectives designed to help further the growth of the ocean renewable energy sector and bring it closer to commercialisation. One of the objectives of the OPERA project is to apply some of the recently developed IEC 62600 standards to an OWC wave energy converter (WEC). The standards applied in the OPERA project include the IEC 62600-30: Electrical power quality requirements for wave, tidal, and other water current energy converters.

As part of the maturation process, the ocean renewable energy sector must prove that it can provide consistent electricity with power quality at a level that is safe and non-disruptive to both private and commercial consumers [7]. Power quality concerns of ocean renewable energy converters have been known for years and models have been built to help identify problems that can arise from grid-connected converters [8,9]. Methods have also been developed and tested to mitigate power quality issues caused by WECs [10]. However, internationally accepted standards for the power quality of WECs have not been tested to date, nor are there many instances where models for renewable energy converters of any time used for power quality assessment are verified against real-world data. The OPERA project offers a unique opportunity to apply IEC power quality standards to sea trials from a grid-connected and fully operational WEC and use the data collected from the sea trials to verify laboratory generated data from a model working with a hardware-in-the-loop (HIL) WEC emulator.

The sea trial data were generated by the Mutriku Wave Power Plant (MWPP), which played an intricate role during the OPERA project, as it was utilised for experimentation and data collection [11,12]. The MWPP is a 175 kW OWC array built into the breakwater in Mutriku, Spain, on the Bay of Biscay [5]. For the application of the IEC 62600-30 standards, the MWPP electrical power take-off (PTO) system was retrofitted with a Supervisory Control and Data Acquisition (SCADA) system to the requirements of the specifications. The SCADA was active for a period of four months, where it collected two datasets per day for further analysis, and those datasets were used to determine harmonic current distortion caused by the OWC array. The laboratory trials were performed at the Lir National Ocean Test Facility (NOTF) within the MaREI Centre at University College Cork (UCC), Ireland using a 25 kW HIL-based WEC emulator [13,14]. The data collection and analysation methods for both the sea trial and laboratory trial testing are presented in this paper, along with the results of the harmonic current distortion analysis.

2. Data Collection and Analysis Methods

The international technical specification IEC 62600-30 focuses on power quality issues and parameters for single-phase and three-phase, grid-connected and off-grid MECs. Poor electrical power quality negatively affects both power sources and loads, so this technical specification was produced to establish measurement methods, application techniques, and results interpretation guidelines to account for the electrical performance of MECs.

The measurement procedures specified in the IEC 62600-30 are valid for a single MEC unit with a three-phase grid or an off-grid connection. They are designed to be as non-site-specific as possible. For analytical purposes, the IEC 62600-30 divides marine renewable energy resources into three classifications: low, medium, and high. For WECs, the resources can be classified through the sea summary statistics using either significant wave height (H_s) or energy period (T_e). The decision of which statistic to use is made by the developer, as is the parsing of the resource classification, though the classifications should be based annual conditions of the deployment site. The procedures are valid for any size of MEC, with varying specifications depending on the type of voltage connection, which includes three classes of connection, low voltage (LV), medium voltage (MV), and high voltage (HV).

In this paper, the IEC 62600-30 guidelines for measuring harmonic distortion were applied to sea trial testing from the MWPP and laboratory testing at the Lir NOTF. The laboratory testing was designed to mimic conditions at the MWPP, as both systems have LV connections to the grid and similar resource conditions. The resource classifications for the trial presented here were based on H_s. Low energy seas are $H_s < 1.25$ m; medium energy seas are 1.25 m $\leq H_s < 2.5$ m; high energy seas are $H_s \geq 2.5$ m. The IEC 62600-30 recommends that at least five datasets under each resource classification are analysed. For the sea trials, 24 datasets, collected over a period of four months from January through April 2013, were analysed, with eight datasets for each resource. For the laboratory trials, 14 datasets were generated of the most common sea states observed annually. The datasets were collected and processed according to the IEC documentations.

2.1. IEC 62600-30 Harmonic Current Analysis Methods

IEC 62600-30 documentation stipulates that the emission of current harmonics needs be measured and recorded. As recommended in IEC 62600-30, the harmonic distortion analysis was based on IEC 61000-4-7: testing and measurement techniques—general guide on harmonics and interharmonics measurements and instrumentation for power supply systems. The harmonic distortion analysis had three separate determinations: harmonic distortion, interharmonic distortion, and high-frequency harmonic distortion. The analysis and observations were based on the fast Fourier transform (FFT) of the original current signal, as outlined in the IEC 61000-4-7 documentation, this includes setting the duration of the time window for the FFT at 200 ms. The FFT of the original signals was performed using the 'fft(x)' function developed by The Mathworks, Inc. for their MATLAB software.

The harmonic distortion refers to signals at frequencies which are an integer multiple of the fundamental frequency of the power system. The power systems evaluated in this report had a fundamental frequency of 50 Hz, so the harmonics of the system were 100 Hz, 150 Hz, 200 Hz, up to and including 2.5 kHz. Interharmonics refers to spectral components with frequencies between two consecutive harmonic frequencies. High-frequency harmonics refers to those signals with frequencies above 2 kHz and below 50% of the sampling frequency. The datasets were generated over a period of 10 minutes. The sampling frequency was 15 kHz for the sea trial datasets, and the sampling frequency was 20 kHz for the laboratory datasets.

2.1.1. Determination of Harmonic Currents Below 2.5 kHz

Harmonic distortion below 2.5 kHz represents the harmonic orders from $h = 2$ to $h = 50$ for a 50 Hz signal, where h is the integer ratio of a harmonic frequency to the fundamental frequency of the power system.

For the assessment of harmonics, the output of the FFT was grouped with the sum of the squared intermediate components between two adjacent harmonics, as shown in Equation (1). The FFT analysis assumed a stationery signal, but the magnitude of power systems tends to fluctuate, spreading out the energy of the harmonic components to adjacent spectral component frequencies [15]. To account for the fluctuations in signal, the output components for each 5 Hz of the FFT were grouped using Equation (1), as given in IEC 61000-4-7. The resulting harmonic subgroup of order h has a magnitude:

$$I_{sg,h}^2 = \sum_{k=-1}^{1} I_{C,(N \times h)+k'}^2 \tag{1}$$

where $I_{sg,h}$ is the resulting Root Mean-Square (RMS) current value of the harmonic subgroup, h is the integer multiple of the fundamental frequency that represents the harmonic order, $I^2_{C,(N \times h) + k}$ is the RMS value of the spectral component corresponding to an output bin of the FFT, N is the number of power sully periods within the window, h is the harmonic group frequency order, and k is the order of the spectral components. The output components for each 5 Hz of the FFT were grouped by Equation (1) as shown in Figure 1 to improve the assessment accuracy of the current, as directed by IEC 61000-4-7 [15].

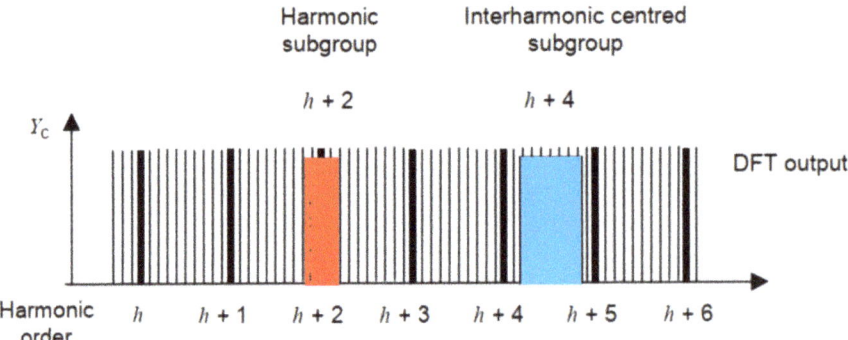

Figure 1. Illustration of the harmonic subgroups in red and interharmonic centred subgroups in blue [15].

The resulting RMSs of the amplitude of the current harmonic subgroups were used to determine the harmonic current distortion for each harmonic from $h = 2$ to $h = 50$. IEC 62600-30 states that harmonic currents below 0.1% of the device rated current, I_r, for any of the harmonic orders need not be reported. Equation (2) was used to determine the normalised currents of a given harmonic order and if those currents need to be reported:

$$\left(\frac{I_{sg,h}}{I_r}\right)\% = \frac{\sqrt{\sum_{k=-1}^{1} I^2_{C,(N\times h)+k}}}{I_r \sqrt{2}} * 100, \tag{2}$$

where I_r is the rated current of the wave energy converter. Again, the output components for each 5 Hz of the FFT were grouped by Equation (1), as shown in Figure 1.

2.1.2. Determination of Interharmonic Currents Below 2.5 kHz

Interharmonics below 2.5 kHz represent the RMS values of current components whose frequencies are not an integer of the fundamental, which appear as discrete frequencies of a wide-band spectrum [16]. A grouping of the spectral components in the interval between two consecutive harmonic components forms an interharmonic group [15].

Interharmonic components are caused primarily by two sources: variations in the amplitude and/or phase angle of the fundamental component and/or of the harmonic components, and power electronics circuits with switching frequencies not synchronised to the power supply frequency and power factors correctors. Potential effects include additional torques on motors and generators, disturbed zero crossing detectors, and additional noise in inductive coils.

Spectral components associated with interharmonics usually vary in both magnitude and frequency. Grouping them provides an overall value for the spectral components between two discrete harmonics, which includes the effects of fluctuations of the interharmonic components. To further reduce the effects of amplitude and phase angle fluctuations, components immediately adjacent to the to the harmonic frequencies that the interharmonics were between were excluded by using Equation (3), which is given in Annex A of 61000-4-7:

$$I^2_{isg,h} = \sum_{k=2}^{N-2} I^2_{C,(N\times h)+k'} \tag{3}$$

where $I^2_{C,(N\times h)+k}$ is the RMS value of the spectral component corresponding to an output bin of the FFT that exceeds the frequency of the harmonic order h, $I^2_{isg,h}$ is the RMS value of the interharmonic current centred subgroup of order h.

The resulting RMSs of the amplitude of the current interharmonic subgroups were used to determine the interharmonic current distortion between harmonics from $h = 2$ to $h = 40$. Equation (4) was used to determine if the normalised currents of a given harmonic order need to be reported:

$$\left(\frac{I_{isg,h}}{I_r}\right)\% = \frac{\sqrt{\sum_{k=2}^{N-2} I_{C,(N \times h)+k}^2}}{I_r \sqrt{2}} * 100, \tag{4}$$

2.1.3. Determination of High-Frequency Harmonic Currents

High-frequency harmonics are components in signals with frequencies above the 40th harmonic, which is 2 kHz for a 50 Hz system. They can be caused by several phenomena, including Pulse-Width Modulation (PWM) control of power supplies at the mains side connection, emissions like mains signalling, feed-through from the load or generator side of the power converters to the mains system side, and oscillations due to commutation notches. The measurement of these components is grouped into predefined frequency bands based on the signal energy of each band.

The FFT output was grouped into 200 Hz bands, beginning at the first centre band above the harmonics range. For the analysis of 50 Hz signals, the first centre band frequency was 2100 Hz. The RMSs of the amplitude of the high-frequency current bands were used to determine the harmonic current distortion between harmonics from $f = 2100$ Hz to $f = 7500$ Hz for the sea trials and $f = 2100$ Hz to $f = 8900$ Hz for the laboratory trials. Equation (5) was used to determine if the normalised currents of a given harmonic order need to be reported:

$$\left(\frac{I_{hfh,f}}{I_r}\right)\% = \frac{\sqrt{\sum_{k=b-95}^{b+100} I_{C,f}^2}}{I_r \sqrt{2}} * 100, \tag{5}$$

where $I_{hfh,f}$ is the current amplitude at the frequency of f, I_r is the rated current of the marine energy converter, and $I_{C,f}$ is the RMS value of the current component C.

2.2. Mutriku Wave Power Plant and SCADA System

The MWPP is composed of 16 fixed-type OWCs in an array. Each of the 16 OWC chambers has a two-stage Wells turbine and an 18.5 kW electrical generator that acts as the power take-off (PTO) system. The power generated by the wave energy plant is supplied to the local grid and accounts for approximately 400 MWh of carbon-free electricity annually [5].

Each OWC generator is controlled by an individual variable frequency drive (VFD), which allows the turbines to operate efficiently over a wide range of sea state conditions. The VFDs feed into a DC-bus, which is one of two within the plant. A single DC-bus accounts for eight generators, and each DC-bus supplies the grid though a 115 kW DC-AC converter that uses a VFD to sync to the AC output frequency with the local grid. The voltage and current outputs of a single DC-AC converter were monitored for the application of the IEC 62600-30 standards.

The SCADA system used to apply the IEC 62600-30 standards to the MWPP was connected to one of the 115 kW converters used to supply the grid. Measurements were taken from between the DC-AC converter and the radio frequency interference (RFI) filter that separated the grid from the converter. Voltage and current transducers installed within the plant's electrical system transformed the voltages and currents into signals that voltages that could be monitored by the SCADA system.

The main processing unit of the SCADA was a National Instruments (NI) cRIO-9082 operating NI Labview software. The cRIO-9082 is an eight-slot cRIO with a 1.33 GHz dual-core CPU, 2 GB of DRAM, 32 GB of ROM, and a Xilinx Spartan-6 LX150 FPGA. The cRIO was populated with NI-9239 analogue input modules with a voltage measuring range of −10 to 10 Volts. The NI-9239 has a sampling rate up to 50 kHz.

These specifications were vital for applying the IEC 62600-30 standards. To perform the harmonic analysis presented in this report, the IEC 62600-30 required a 10-minute continuous dataset sampled at 20 kHz. The Xilinx Spartan-6 LX150 FPGA along with the NI-9239 analogue import cards allowed for high-frequency sampling, but the sampling frequency was limited to 15 kHz, which is below the 20 kHz stipulated in IEC 62600-30. This limitation was due to the card being responsible for monitoring three signals. The number of signals handled by a single card affects the sampling rate, and for three signals, the maximum sampling rate was 16.67 kHz. The high-frequency harmonic analysis was limited to 7.5 kHz, rather than 10 kHz, because of the lower sampling frequency applied to the dataset. The sampling frequency and the 10-minute duration required required over 1 GB of memory per dataset. During deployment of the SCADA system, the 32 GB ROM had to be routinely cleared, with each dataset being moved to the cloud-based data storage system.

As the NI-9239 analogue input cards had a voltage range of ±10 V, voltage and current transducers were installed as part of the SCADA system to convert the voltages and currents to signals that could be monitored by the analogue input cards. LEM supplied both the transducers for the voltage and current measurement to ensure continuity in the datasets. The voltage transducers were LEM DVL 750, which provide bipolar and insulated measurement up to 1125 V. The output of the transducers was a milliamp current with a set mA/V ratio to represent the measured voltage. The current transducers were LEM LA 305-S, which are Hall-effect closed-loop transducers, with a maximum range of ±500 A. The output of the transducers was a milliamp current with a set mA/A ratio to represent the measured current. To produce signals that could be perceived by the cRIO, high tolerance resistors were placed in series with the current outputs of the transducers, as specified by the LEM provided technical data sheets for both transducers. Figure 2 shows the voltages and current transducers installed at the Mutriku Wave Power Plant; at the time of the photos, the voltage signals had not yet been wired to the voltage transducers.

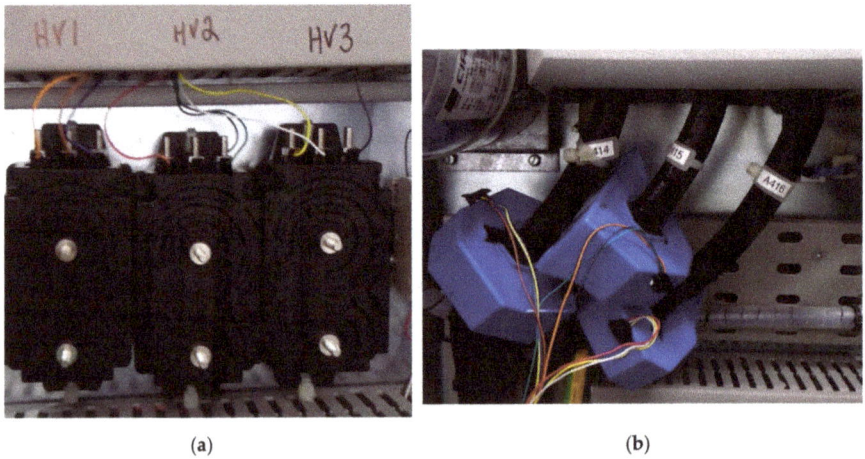

Figure 2. The voltage and current transducers installed in the Mutriku Wave Power Plant: (**a**) three LEM DVL 750 voltage transducers; (**b**) three LEM LA 305-S current transducers.

The voltage and current transducers were retrofitted into the already operational MWPP, and there were difficulties with the physical placement of the current transducers due to space and wiring requirements. Ideally, the voltage and current measurements for the application of IEC 62600-30 should be taken after the RFI filter that was used to remove the high-frequency signal generated by the VFD in renewable energy generation systems. Unfortunately, the space and wiring requirements forced the installation of the transducers between the RFI filter and the grid-side VFD, and this influenced the results of the testing. Figure 3 is a single-line diagram for an individual DC-bus of the plant, which shows the ideal placement of the voltage and current transducers in green and the actual placement of the voltage and current transducers in red.

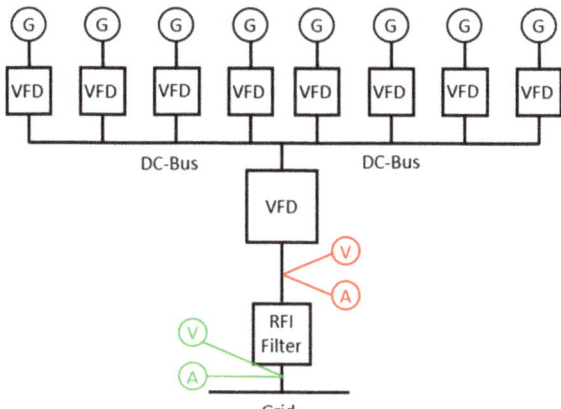

Figure 3. Mutriku wave power plant single line drawing showing how the eight generators were connected to the grid via a single DC-bus and the ideal and actual placement of the transducers.

2.3. Lir National Ocean Test Facility Electrical Lab and SCADA System

The electrical laboratory at the Lir NOTF at UCC includes a medium speed rotary emulator. The rotary emulator, shown in Figure 4, is an electromechanical system used to mimic the rotating electrical power take-off (PTO) system of a renewable energy device. The emulator is composed of two electrical machines directly coupled by a mechanical shaft, with a torque transducer between them. The mechanical drive shaft also includes a stainless-steel flywheel that is connected to the system by a five-position gear box. The flywheel allows the drive shaft to be composed of one of five different inertial masses, which can be implemented to replicate the inertia of any system being tested. The prime mover, which is used to emulate the forces applied by the turbine, is a four-pole squirrel-cage induction machine (SCIM) with a rated power of 22 kW, a rated speed of 1467 rpm, and a rated torque of 143 Nm. The generator is a slip ring four-pole induction machine with a rated power of 22 kW, a rated speed of 1472 rpm, and a rated torque of 143 Nm. As described in [13], the generator rotor can be set in multiple configurations depending on the system which is being emulated. For the OPERA project, the rotor is configured as an SCIM.

The emulator includes HIL technology, allowing it to operate in real-time in conjunction with complex software modelling running on the MATLAB® Simulink® platform. The conditions modelled in the Simulink® software are relayed to the emulator to drive its behaviour, and the conditions of the physical testing equipment are fed back into the Simulink® model to affect the model and complete the loop. For the MWPP, the data sent to the emulator by the Simulink® model were turbine torque and chamber pressure, while the feedback from the emulator to the model was turbine speed. The emulator–HIL integration was meticulously characterised and verified to ensure accuracy of the software–hardware link [14].

As part of the OPERA project, the MWPP was modelled using Simulink® [12]. This model was combined with the HIL system at the Lir NOTF to facilitate the laboratory testing. During operation of UCC's emulator and model, the turbine input torque determined in Simulink® is sent to the Programmable Logic Controller (PLC) as a reference torque to drive the motor that acts as the turbine for laboratory testing. The controller for the generator resides on the PLC and determines the generator braking torque to extract power from the turbine and export electrical power to the grid.

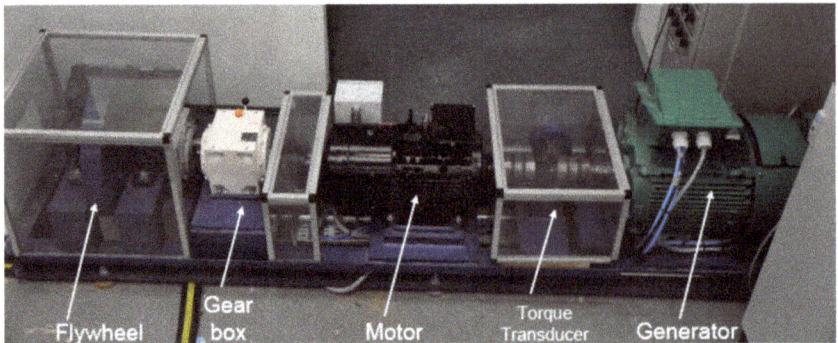

Figure 4. Drive train elements of the medium speed rotary emulator at the Lir NOTF laboratory.

The power rating of the turbine–generator PTO system installed at the MWPP was 30 kW, while the power rating of the turbine emulator and the generator of the HIL system was 22 kW. Due to this discrepancy, the model used for HIL testing had to be scaled down slightly to allow for testing. As is common in wave energy, the Froude scaling method was applied to the HIL model, with a scaling factor of 1:1.25.

The Lir NOTF electrical laboratory includes an integrated SCADA system that is very similar to the system installed at the MWPP, allowing for the laboratory generated datasets to be compared with the datasets from the sea trial testing. The Lir NOTF SCADA system also uses an NI cRIO-9082 running NI LabVIEW software for high-frequency data collection. The cRIO included NI-9225 cards rated for 300 V-RMS and NI-9239 cards with a voltage measuring range of −10 to 10 Volts. The NI-9225 cards were used to measure the line-to-neutral voltage at the point of common coupling (PCC) between the grid and the DC-AC converter output of the medium speed rotary emulator. The NI-9239 cards monitored the output of the current transducers for each of the three phases of the converter. The current transducers were LEM LA 55-P, with a current range of 0 to 50 A. The output of the current transducers was analogue current signals with a step-down ratio of 1000:1. The cRIO was fit with high tolerance resistors rated to LEM specification to produce a voltage that could be monitored by the cRIO NI-9239 cards. The SCADA system in the lab allowed for the data to be collected at the recommended sampling frequency of 20 kHz for the 10-minute window tests. The voltage and current transducers in the Lir NOTF were able to be installed after the RFI filter rather than before the filter. Figure 5 shows a single-line diagram of the laboratory arrangement, showing the placement of the current and voltage meters within the Lir NOTF electrical laboratory.

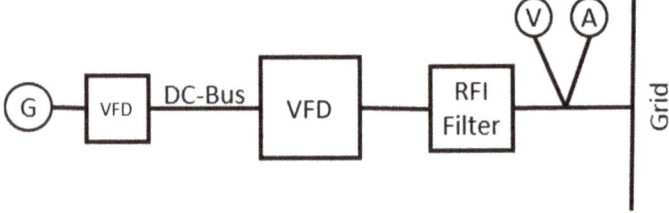

Figure 5. Single line drawing of the Lir NOTF emulator showing the placement of the voltage and current transducers.

3. Results of Harmonic Analysis for Sea Trial and Laboratory Generated Datasets

There were 24 datasets from the sea trials and 14 datasets from the laboratory testing that were analysed for current harmonic distortion. As the laboratory tests had to be modestly scaled and the power output of the grid side VFDs was different, the current data were normalised against rated current for comparison. The datasets were spread across the three resource classifications, as identified in Section 2. Table 1 shows the significant wave height and energy period for both the sea trial and laboratory trial datasets, including the resource classifications. Although each dataset was processed and analysed individually, the results presented in this section represent the average values for each resource classification.

Table 1. Summary sea-state statistics for the datasets analyzed in the paper.

	Sea Trials			Laboratory Trials	
Class	H_s (m)	T_e (s)	Class	H_s (m)	T_e (s)
Low	0.50	17.80	Low	0.88	5.50
	0.54	8.06		1.03	6.50
	0.69	10.81		1.04	7.50
	0.74	9.98		1.02	8.50
	0.90	13.03		1.08	9.50
	0.94	13.34		1.19	10.50
	1.00	11.55		1.48	11.50
	1.09	10.12	Medium	1.81	12.50
	1.33	12.53		2.07	13.50
	1.39	16.43		2.11	12.50
Medium	1.58	15.51		2.59	14.50
	1.86	13.61	High	2.88	15.50
	1.93	15.05		3.16	16.50
	2.11	12.89		3.20	11.50
	2.23	14.33			
	2.41	13.85			
	2.62	14.31			
	2.64	13.65			
High	2.66	13.28			
	2.74	15.54			
	3.22	16.19			
	3.35	15.65			
	3.69	15.68			
	4.14	16.81			

3.1. Measured Harmonic Currents Below 2.5 kHz

3.1.1. Sea Trial Datasets

As defined in Section 2.1.1., harmonic distortion below 2.5 kHz represents the harmonic orders from $h = 2$ to $h = 50$ for a 50 Hz signal, where h is the integer ratio of a harmonic frequency to the fundamental frequency of the power system. The harmonic current subgroups were identified via FFT. The RMS of the amplitude of the current harmonic subgroups was used to determine the harmonic current distortion for each harmonic from $h = 2$ to $h = 50$.

The harmonic current amplitude for the datasets generated during the sea trial was processed using the same FFT software referenced in Section 2.1. Table 2 shows the average harmonic currents as a percentage of the rated current of the converter, I_r, that reached the 0.1% of I_r reporting threshold for the three resource classifications determined from the collected data based on Equation (2). Figure 6 shows a graphical representation of Table 1, including those harmonic currents below the 0.1% reporting threshold.

Table 2. Sea trial average reportable ratios of harmonic current amplitudes for each resource classification.

	Average Harmonic Current Amplitude/Rated Current (I_h/I_r %)										
h	Sea State			h	Sea State			h	Sea State		
	Low	Medium	High		Low	Medium	High		Low	Medium	High
2	0.337%	0.374%	0.390%	13	0.336%	0.296%	0.273%	22	0.132%	0.138%	0.132%
3	2.713%	2.691%	2.819%	14	0.169%	0.161%	0.171%	23	0.111%	0.100%	0.098%
4	0.266%	0.281%	0.301%	15	0.469%	0.505%	0.601%	24	0.107%	0.112%	0.111%
5	1.581%	1.727%	2.044%	16	0.241%	0.240%	0.253%	33	0.120%	0.115%	0.120%
6	0.201%	0.202%	0.200%	17	0.338%	0.298%	0.312%	39	0.221%	0.238%	0.232%
7	0.401%	0.420%	0.457%	18	0.134%	0.132%	0.130%	41	0.218%	0.235%	0.228%
9	0.106%	0.097%	0.108%	19	0.223%	0.192%	0.193%	46	0.098%	0.093%	0.102%
11	0.305%	0.366%	0.339%	20	0.148%	0.152%	0.146%	50	0.277%	0.247%	0.267%

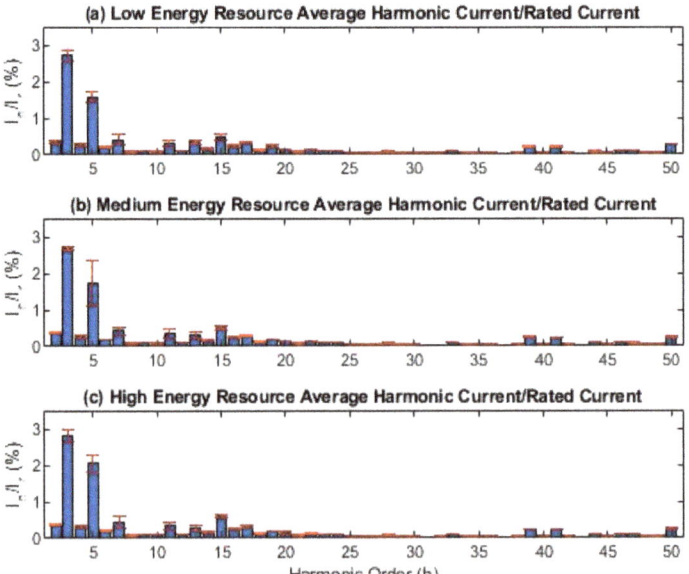

Figure 6. Sea trial average ratios of harmonic current amplitude to rated current for (**a**) low energy, (**b**) medium energy, and (**c**) high energy resource classifications, including error bars, which represent the standard deviation of the data averaged.

3.1.2. Laboratory Trial Datasets

The harmonic current amplitude for the datasets generated during laboratory testing was processed using the same FFT software that was applied to the sea trial data. Table 3 shows the average harmonic currents as a percentage of I_r, for the three resource classifications determined from the collected data, based on Equation (2). Figure 7 is a graphical representation of Table 3, including those harmonic currents below the 0.1% reporting threshold.

The most significant harmonic currents observed during the laboratory testing occurred at the fifth harmonic, 250 Hz, with additional notable harmonic currents occurring at the third harmonic, 250 Hz. The amplitudes of the remaining harmonic currents were below 1% of I_r with a number of them low enough that they need not be reported. For verification purposes, other notable peaks, those above 0.4% of I_r, in harmonic currents occurred at the 7th, 15th, 17th, 19th, 20th, and 26th harmonics. At harmonic orders above the 26th, only the 32nd, 37th, 38th, and 43rd harmonics reached the reporting threshold. As with the sea trial datasets, there was little variation in the observed harmonic currents across the three resource classifications, with small increases seen in the larger harmonic current amplitudes in the high energy classification. The amplitudes of the third and fifth harmonics exhibited any noticeable changes with changes in available energy in the sea condition, with the largest observed harmonic current for either trial occurring in the fifth order harmonic of the laboratory trial datasets.

Table 3. Sea trial average reportable ratios of harmonic current amplitudes for each resource classification.

	Average Percentage Harmonic Current Amplitude/Rated Current (I_h/I_n %)										
h	Sea State			h	Sea State			h	Sea State		
	Low	Medium	High		Low	Medium	High		Low	Medium	High
2	0.322%	0.379%	0.421%	13	0.272%	0.315%	0.265%	22	0.066%	0.071%	0.071%
3	1.962%	2.138%	2.073%	14	0.491%	0.465%	0.480%	23	0.175%	0.262%	0.183%
4	0.336%	0.402%	0.452%	15	0.633%	0.633%	0.640%	24	0.048%	0.051%	0.053%
5	2.868%	2.658%	3.155%	16	0.129%	0.138%	0.150%	33	0.031%	0.032%	0.033%
6	0.203%	0.219%	0.249%	17	0.608%	0.485%	0.682%	39	0.032%	0.030%	0.032%
7	0.960%	0.766%	0.729%	18	0.151%	0.167%	0.189%	41	0.037%	0.037%	0.035%
9	0.365%	0.335%	0.310%	19	0.512%	0.594%	0.734%	46	0.029%	0.027%	0.027%
11	0.388%	0.462%	0.485%	20	0.547%	0.543%	0.537%	50	0.070%	0.074%	0.071%

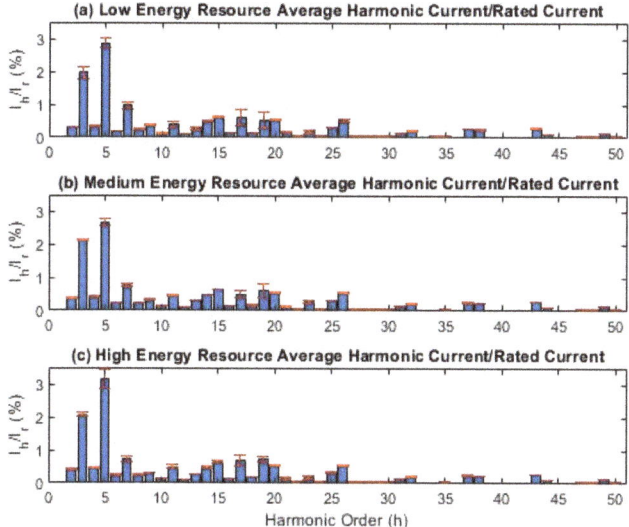

Figure 7. Laboratory trials average ratios of harmonic current amplitude to rated current for (**a**) low energy, (**b**) medium energy, and (**c**) high energy resource classifications, including error bars, which represent the standard deviation of the data averaged.

3.2. Determination of Interharmonic Currents Below 2.5 kHz

3.2.1. Sea trial Datasets

As defined in Section 2.1.2., interharmonics below 2.5 kHz represent the current RMS values of current components whose frequencies are not an integer of the fundamental, which appear as discrete frequencies of as a wide-band spectrum. A grouping of the spectral components in the interval between two consecutive harmonic components forms an interharmonic group, and the values of the interharmonic group current amplitude from the sea trials are presented in this section.

The groupings provide an overall value for the spectral components between two discrete harmonics, which includes the effects of fluctuations of the interharmonic components. To further reduce the effects of amplitude and phase angle fluctuations, components immediately adjacent to the harmonic frequencies that the interharmonics are between are excluded. The resulting RMS of the amplitude of the current interharmonic subgroups are used to determine the interharmonic current distortion between harmonics from $h = 2$ to $h = 40$. The interharmonic current amplitudes were determined with the same FFT software that was used to determine the harmonic currents.

Shows the average interharmonic currents as a percentage of I_r, for the three resource classifications determined from the collected data based on Equation (4). Figure 8 is a graphical representation of Table 4

Table 4. Sea trials average reportable ratios of harmonic current amplitudes for each resource classification.

	Average Percentage Interharmonic Current Amplitude/Rated Current (I_{ih}/I_r %)										
ih	Sea State			ih	Sea State			ih	Sea State		
	Low	Medium	High		Low	Medium	High		Low	Medium	High
1	0.444%	0.684%	0.676%	14	0.173%	0.167%	0.173%	27	0.060%	0.063%	0.063%
2	0.325%	0.379%	0.377%	15	0.216%	0.220%	0.238%	28	0.048%	0.047%	0.048%
3	0.322%	0.346%	0.344%	16	0.128%	0.142%	0.148%	29	0.049%	0.050%	0.049%
4	0.240%	0.259%	0.264%	17	0.157%	0.118%	0.116%	30	0.037%	0.037%	0.039%
5	0.184%	0.194%	0.194%	18	0.102%	0.094%	0.095%	31	0.038%	0.037%	0.038%
6	0.136%	0.145%	0.150%	19	0.129%	0.129%	0.124%	32	0.060%	0.057%	0.054%
7	0.110%	0.116%	0.119%	20	0.076%	0.076%	0.076%	33	0.064%	0.063%	0.064%
8	0.083%	0.090%	0.093%	21	0.085%	0.085%	0.085%	34	0.037%	0.039%	0.038%
9	0.067%	0.075%	0.077%	22	0.074%	0.073%	0.073%	35	0.047%	0.052%	0.050%
10	0.059%	0.068%	0.067%	23	0.068%	0.066%	0.067%	36	0.038%	0.038%	0.038%
11	0.062%	0.068%	0.069%	24	0.053%	0.053%	0.052%	37	0.053%	0.054%	0.053%
12	0.078%	0.093%	0.082%	25	0.047%	0.047%	0.047%	38	0.101%	0.108%	0.097%
13	0.102%	0.100%	0.104%	26	0.043%	0.043%	0.044%	39	0.134%	0.153%	0.148%

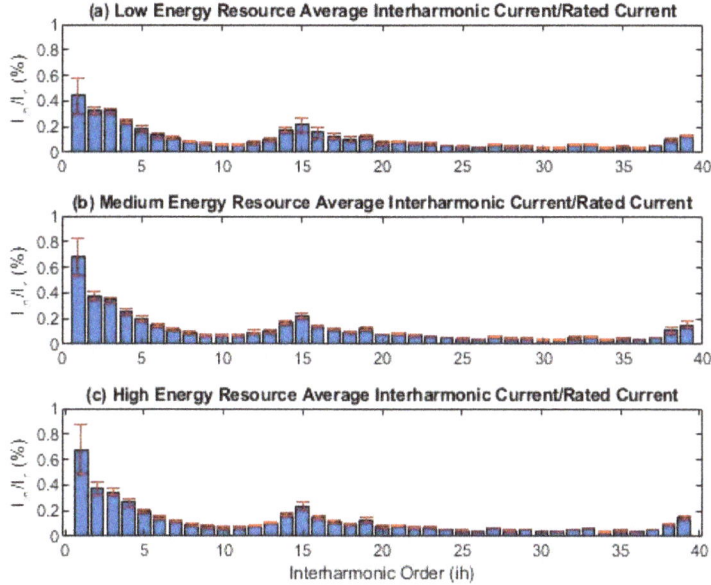

Figure 8. Sea trial average ratio of interharmonic current amplitude to rated current for (**a**) low energy, (**b**) medium energy, and (**c**) high energy resource classifications including errorbars that represent the standard deviation of the data averaged.

For the interharmonics, the most significant harmonic currents occur at the 1st interharmonic, which represents the window around 75 Hz. With the exception of the 1st interharmonic current, there is no discernible change in the interharmonic currents across the three resource classifications. There is a clear distribution across the interharmonic orders with peak currents at the 1st order, reaching a minimum in the lower interharmonics at the 10th order, reaching a secondary maximum at the 15th order, and sitting an overall minimum until the 39th order interharmonic where there is a final small peak in current. There is a 50% increase in the 1st order interharmonic currents observed from the low energy to the medium and high energy classifications, with little change in the observed interharmonic currents beyond the 1st order.

3.2.2. Laboratory Datasets

The interharmonic current amplitudes for the laboratory datasets were processed using the same software that was used for the sea trial datasets. Table 5 shows the average interharmonic currents as a percentage of I_r, for the three resource classifications determined from the collected data based on Equation (5). Figure 9 is a graphical representation of Table 5, which includes those currents that fall below the reporting threshold.

For the interharmonics, the most significant currents occur at the second and third interharmonic, which represents the window around 125 Hz and 175 Hz respectively. Including the below the 20th interharmonic order, there is an increase in the current amplitude of each interharmonic order with increasing energy available in the sea conditions. The increases are very small, with only the current amplitudes of the interharmonic orders first through fourth seeing increases above 0.1% of I_{ih}/I_r between the low energy and high energy conditions. Above the 20th order, there is little change between the sea state energy classifications. The interharmonic currents observed in the laboratory testing datasets were on average larger than those observed in the sea trial datasets. Like the sea trial interharmonics, there is little change in interharmonic currents over the resource classifications.

Table 5. Laboratory trials average reportable ratios of harmonic current amplitudes for each resource classification.

	Average Percentage Interharmonic Current Amplitude/Rated Current (I_{ih}/I_n %)										
ih	Sea State			ih	Sea State			ih	Sea State		
	Low	Medium	High		Low	Medium	High		Low	Medium	High
1	0.432%	0.493%	0.519%	14	0.158%	0.169%	0.180%	27	0.048%	0.039%	0.041%
2	0.540%	0.567%	0.681%	15	0.154%	0.161%	0.183%	28	0.041%	0.035%	0.040%
3	0.543%	0.623%	0.668%	16	0.162%	0.178%	0.204%	29	0.036%	0.033%	0.035%
4	0.424%	0.500%	0.630%	17	0.170%	0.180%	0.231%	30	0.050%	0.044%	0.044%
5	0.343%	0.364%	0.405%	18	0.145%	0.162%	0.187%	31	0.064%	0.069%	0.069%
6	0.283%	0.302%	0.335%	19	0.123%	0.137%	0.153%	32	0.036%	0.042%	0.050%
7	0.245%	0.258%	0.284%	20	0.103%	0.113%	0.123%	33	0.033%	0.027%	0.030%
8	0.210%	0.220%	0.249%	21	0.071%	0.060%	0.057%	34	0.034%	0.030%	0.035%
9	0.184%	0.191%	0.210%	22	0.058%	0.063%	0.060%	35	0.029%	0.027%	0.030%
10	0.165%	0.173%	0.189%	23	0.047%	0.050%	0.051%	36	0.053%	0.050%	0.056%
11	0.149%	0.157%	0.175%	24	0.075%	0.068%	0.071%	37	0.076%	0.082%	0.088%
12	0.142%	0.148%	0.159%	25	0.133%	0.149%	0.152%	38	0.046%	0.045%	0.047%
13	0.175%	0.181%	0.193%	26	0.083%	0.093%	0.101%	39	0.030%	0.027%	0.028%

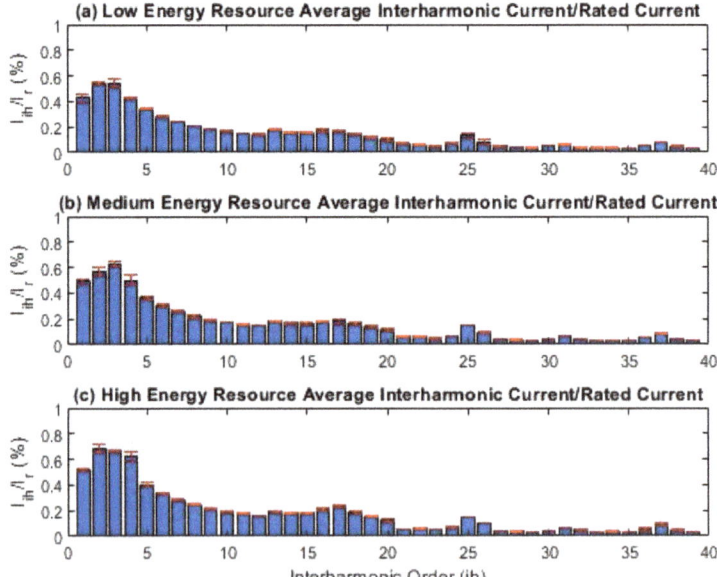

Figure 9. Laboratory trials average ratio of interharmonic current amplitude to rated current for (**a**) low energy, (**b**) medium energy, and (**c**) high energy resource classifications including errorbars that represent the standard deviation of the data averaged.

3.3. Determination of High-frequency Harmonic Currents

3.3.1. Sea trial Datasets

As defined in Section 2.1.3, high-frequency harmonic are components in signals with frequencies above the 40th harmonic, which is 2 kHz for a 50 Hz system. The measurement of high-frequency harmonic components is grouped into predefined frequency bands based on the signal energy of each band according to IEC 61000-4-7.

The FFT output is grouped into 200 Hz bands beginning at the first centre band above the harmonics range. For the analysis of 50 Hz signals, the first centre band frequency is 2100 Hz. The RMS of the amplitude of the high-frequency current bands are used to determine the harmonic current distortion between harmonics from f = 2100 Hz to f = 7500 Hz for the sea trial datasets because the sampling frequency applied to the datasets was limited to 15 kHz, which would cause aliasing for frequencies above 7500 Hz.

The average high-frequency harmonic currents identified in the sea trial datasets for the three resource classifications are presented in Table 6 as a percentage of I_r, which is determined from the collected data based on Equation (5). Figure 10 is a graphical representation of Table 6.

The high-frequency signals prevalent in the data collected from the sea trials are directly related to the PWM switching of the grid side VFD used to create a 50 Hz sine wave to deliver power from the WEC to the grid, which has a switching frequency of 3 kHz. The largest currents are around the 3 kHz switching, with secondary currents at 6 kHz, which is the 2nd harmonic of 3 kHz. All other high-frequency harmonics currents are well under 1% of I_r. The current amplitudes observed at 2900 Hz and 3100 Hz represent the largest occurring harmonic current amplitudes for the sea trial datasets. This is largely influenced by the switching frequency of the VFD supplying power to the grid and the placement of the current transducers in relation to the RFI filter. There are no changes in the high-frequency harmonic currents across the three resource classifications.

Table 6. Sea trial average reportable ratios of high-frequency harmonic current amplitudes for each resource classification.

	Average Percentage High-frequency Harmonic Current Amplitude/Rated Current (I_{hfh}/I_r %)										
f (Hz)	Sea State			f (Hz)	Sea State			f (Hz)	Sea State		
	Low	Medium	High		Low	Medium	High		Low	Medium	High
2100	0.316%	0.338%	0.331%	4100	0.273%	0.295%	0.283%	6100	0.794%	0.891%	0.828%
2300	0.211%	0.214%	0.222%	4300	0.205%	0.198%	0.204%	6300	0.672%	0.648%	0.657%
2500	0.443%	0.428%	0.446%	4500	0.137%	0.138%	0.135%	6500	0.072%	0.071%	0.074%
2700	2.195%	2.083%	2.225%	4700	0.098%	0.101%	0.098%	6700	0.111%	0.106%	0.108%
2900	4.062%	3.982%	4.095%	4900	0.176%	0.191%	0.178%	6900	0.218%	0.214%	0.222%
3100	4.003%	3.848%	3.959%	5100	0.186%	0.204%	0.193%	7100	0.189%	0.182%	0.187%
3300	1.745%	1.872%	1.982%	5300	0.099%	0.101%	0.099%	7300	0.106%	0.109%	0.115%
3500	0.377%	0.386%	0.398%	5500	0.161%	0.161%	0.161%	7500	0.084%	0.083%	0.085%
3700	0.249%	0.242%	0.251%	5700	0.748%	0.718%	0.734%				
3900	0.252%	0.272%	0.261%	5900	0.974%	1.071%	1.011%				

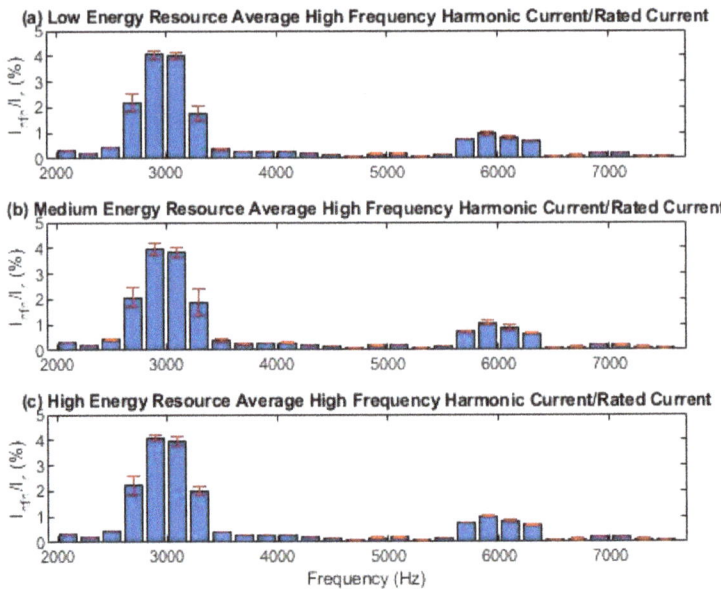

Figure 10. Sea trial average ratios of high-frequency harmonic current amplitude to rated current for (**a**) low energy, (**b**) medium energy, and (**c**) high energy resource classifications, including error bars, which represent the standard deviation of the data averaged.

3.3.2. Laboratory Datasets

The high-frequency current amplitudes for the laboratory datasets were processed using the same software that was used for the sea trial datasets. However, the laboratory datasets were sampled at 20 kHz, which is higher than the 15 kHz frequency used for the sea trial datasets. As a result, the RMSs of the amplitudes of the high-frequency current bands for harmonic current distortion could be determined up to $f = 9000$ Hz.

The average high-frequency harmonic currents identified in the laboratory trial datasets for the three resource classifications are presented in Table 7 as a percentage of I_r, which was determined from the collected data based on Equation (5). Figure 11 is a graphical representation of Table 7.

Like the high-frequency harmonic currents observed in the sea trial datasets, the high-frequency signals prevalent in the data collected in the dry laboratory datasets were directly related to the PWM switching of the grid side VFD used to create a 50 Hz sine wave to deliver power from the MEC to the grid, which had a switching frequency of 3 kHz. The effect of the RFI filter in the laboratory system is evident in these results, which show harmonic currents at the 3 kHz switching frequency to be four-times smaller than those observed in the sea trial datasets. The currents observed around 6 kHz were more similar to those seen in the sea trial datasets. Like the sea trials data, all other high-frequency harmonics currents in the dry laboratory testing were well below 1% of I_r.

Table 7. Laboratory trial average reportable ratios of high-frequency harmonic current amplitudes for each resource classification.

	Average Percentage High-frequency Harmonic Current Amplitude/Rated Current (I_{hfh}/I_r %)										
f (Hz)	Sea State			f (Hz)	Sea State			f (Hz)	Sea State		
	Low	Medium	High		Low	Medium	High		Low	Medium	High
2100	0.325%	0.305%	0.313%	4500	0.225%	0.245%	0.237%	6900	0.103%	0.103%	0.106%
2300	0.114%	0.113%	0.114%	4700	0.140%	0.143%	0.140%	7100	0.144%	0.151%	0.146%
2500	0.191%	0.192%	0.189%	4900	0.350%	0.339%	0.318%	7300	0.092%	0.098%	0.099%
2700	0.279%	0.281%	0.271%	5100	0.256%	0.290%	0.282%	7500	0.157%	0.160%	0.156%
2900	1.075%	1.049%	1.013%	5300	0.170%	0.158%	0.160%	7700	0.091%	0.094%	0.092%
3100	0.746%	0.818%	0.802%	5500	0.529%	0.533%	0.533%	7900	0.142%	0.140%	0.142%
3300	0.281%	0.293%	0.283%	5700	0.147%	0.153%	0.160%	8100	0.146%	0.141%	0.143%
3500	0.304%	0.311%	0.310%	5900	1.654%	1.663%	1.654%	8300	0.068%	0.064%	0.060%
3700	0.124%	0.125%	0.123%	6100	1.305%	1.300%	1.280%	8500	0.095%	0.097%	0.096%
3900	0.381%	0.381%	0.377%	6300	0.143%	0.138%	0.137%	8700	0.200%	0.210%	0.210%
4100	0.413%	0.415%	0.413%	6500	0.354%	0.381%	0.368%	8900	0.304%	0.289%	0.267%
4300	0.155%	0.151%	0.144%	6700	0.198%	0.185%	0.184%				

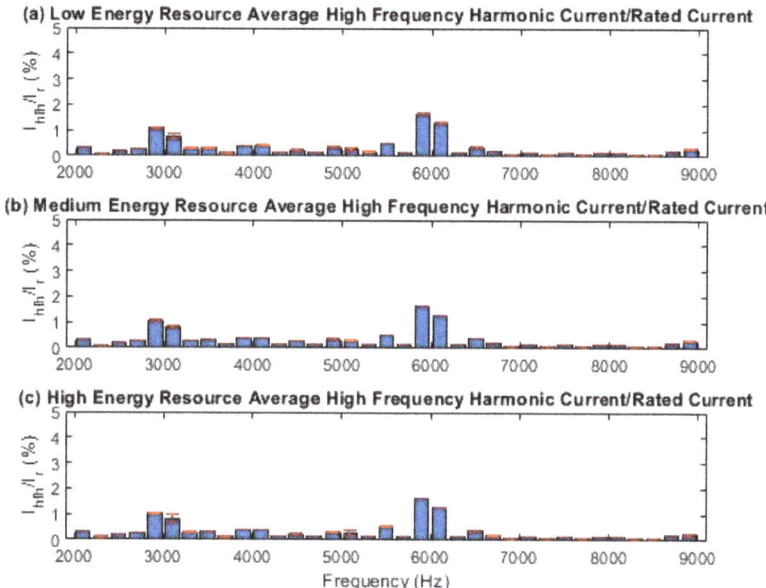

Figure 11. Laboratory trial average ratios of high-frequency harmonic current amplitude to rated current for (**a**) low energy, (**b**) medium energy, and (**c**) high energy resource classifications, including error bars, which represent the standard deviation of the data averaged.

4. Comparison of Sea trial and Laboratory Dataset Findings

To further investigate the validity of the results presented in Section 3, the harmonic, interharmonic, and high-frequency harmonic currents from the laboratory trials were compared against those determined from the sea trials performed at the MWPP. The data presented here represent the average of all the datasets for both the sea trial data and the laboratory generated data.

Figure 12 shows the harmonic, interharmonic, and high-frequency harmonic current amplitudes identified during testing and analysis, which includes data from sea-trails and laboratory trials side by side.

The most significant observations made from the data are that the general scale of the harmonic currents is very similar for both the sea trial and the laboratory generated data, with all harmonic currents falling below 5% of the rated current of the device. The observed harmonic currents were not always identical between sea trial and laboratory data, such as the third order harmonic being more prevalent in the sea trial data, while the fifth order was more prevalent in the laboratory data, but no two grid connections were identical.

The interharmonic currents fell below 1% of the rated current, with very similar results for both datasets. Again, the datasets were not identical, but the pattern of peaks at the lowest interharmonic orders, with a lull in currents around the tenth order before increasing again slightly around the 15th order, was apparent in both datasets.

The largest deviations in the two datasets could be seen at the high-frequency harmonics around 3 kHz. These deviations were directly related to the placement of the current and voltage transducers for the separate experiments. For the laboratory generated datasets, the current transducers were beyond the RFI filter, which was specifically designed to attenuate the 3 kHz switching frequency signal from the VFD, while the sea trial current transducers were located before the RFI filter, which led to the high harmonic current amplitudes observed around 3 kHz in the sea trial testing compared to the dry laboratory datasets. This indicates that the results from both tests are trustworthy and using HIL-based emulators can be used in early device development to evaluate potential harmonic distortion cause by MECs.

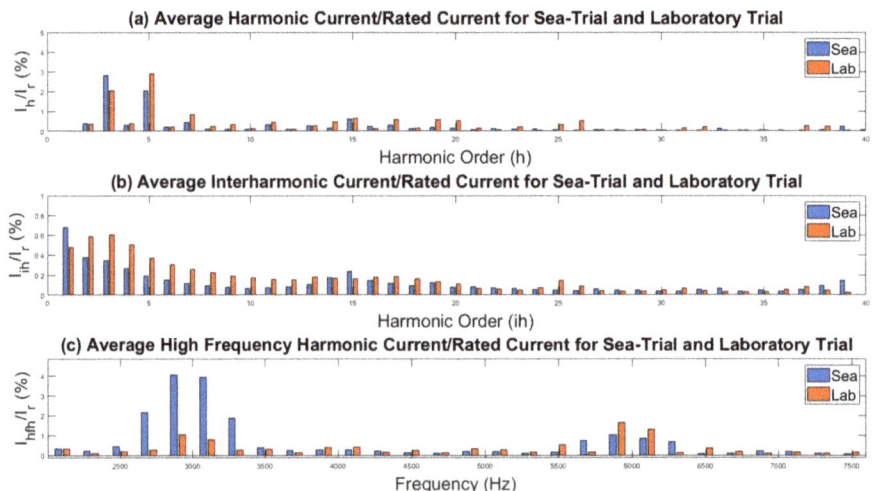

Figure 12. Average normalised harmonic currents observed during sea trial and laboratory trial testing.

Impact of RFI Filter on Data Measurements

The placement of the transducers in relation to the RFI filter for data collection at the MWPP affected the implementation of the IEC 62600-30 standards at the plant and the comparison of the plant datasets with the laboratory datasets. With the current and voltage transducers placed between the VFD and the RFI filter rather than on the grid side of the filter, the datasets were excessively influenced by the pulse-width modulation (PWM) switching of the VFD.

The RFI filters for use in renewable energy systems are specifically designed to attenuate high-frequency noise in signals caused by the PWM switching of VFDs use to generate 50 Hz AC signals from a DC-bus. Figure 13 shows the current measurements taken from the MWPP SCADA system alongside the measurements from the emulator from the electrical laboratory at the Lir NOTF. The measurements from the Lir NOTF were taken from the output of an HIL WEC emulator, which is rated for 25 kW and features a back-to-back AC-DC/DC-AC converter similar to that installed at the MWPP. However, the laboratory generated data were measured by transducers installed on the grid side of the RFI filter rather than the VFD side of the filter. The SCADA system at the Lir NOTF uses the same NI and LEM equipment as was installed at the MWPP.

The difference between the two signals is mostly related to suppression of high-frequency noise by the RFI filter. The sea trial dataset had a more prevalent 3 kHz signal, but the 3 kHz signal could also be clearly observed in the dry laboratory generated dataset. This is reinforced by the results presented in Figure 12, where the largest discrepancies between the two datasets are the harmonic current amplitudes around the 3 kHz switching frequency.

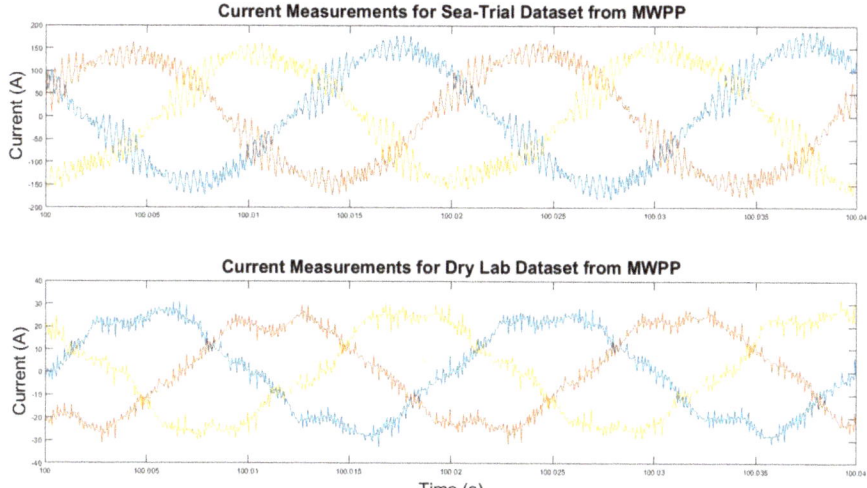

Figure 13. Current measurements from Mutriku Wave Power Plant (MWPP) sea-trails and Lir NOTF laboratory testing.

5. Conclusions

The research presented in this paper represents one of the first attempts to apply the IEC 62600-30 electrical power quality requirements for wave, tidal, and other water current energy converters standards to sea trials of a fully operational wave energy converter, while also using the data collected from the sea trials to attempt to evaluate the accuracy of using laboratory equipment to recreate real sea data. The harmonic distortion of the current created by an OWC was analysed across a range of sea conditions. The data were used to verify an OWC hardware-in-the-loop emulator, which was then used to expand the dataset. Successfully applying and meeting internationally accepted standards is an important step towards commercialisation for the wave energy industry. Several positive conclusions were drawn from the data analysed for this paper.

The IEC 62600-30 standards can be successfully applied to wave energy converters and can help provide confidence in the wave energy sector as it continues to move toward commercialisation.

(a) Small scale oscillating water column wave energy converters can produce power that meets and exceeds internationally accepted standards for harmonic current distortion;
(b) Sea state conditions represented by significant wave height and energy period have little effect on the harmonic currents produced by an oscillating water column wave energy converter.

As the overall energy available in the waves increases, the power injected into the grid increases. However, the corresponding increase in harmonic currents was modest;

(c) While the differences between the sea trial and laboratory trial experimental setup limit the authority of the varication process, the early indication is that laboratory tests can accurately represent harmonic current distortion of sea trial testing. Further validation with more accurately replicated testing procedures needs to take place to confirm these initial results;

(d) The values of the high-frequency harmonic currents around 3 kHz, which are caused by the PWM switching of the VFDs, highlight the importance of including an RFI filter in the development of the electrical power take-off for a wave energy converter.

Author Contributions: Conceptualization, Formal Analysis, Methodology, Validation, Visualization, J.K.; Data curation, E.A.; Project administration, Funding acquisition, P.R.M.; Investigation, Resources, Software, J.K. and E.A.; Supervision, J.K. and P.R.M.; writing – original draft preparation, J.K.; writing – review and editing, E.A. and P.R.M.

Funding: This research has received funding from the European Union's Horizon 2020 research and innovation program under grant agreement No. 654444 (OPERA project).

Acknowledgments: The authors are grateful to the European commission for funding the OPERA projects as part of the Horizon 2020 framework. The authors also thankful to the Basque Energy Board (EVE) for providing technical information of Mutriku Wave Power Plant.

Conflicts of Interest: The authors declare no conflict of interest. The funders had no role in the design of the study; in the collection, analyses, or interpretation of data; in the writing of the manuscript, or in the decision to publish the results.

References

1. International Renewable Energy Agency (IRENA). *Renewable Energy Statistics 2018*; International Renewable Energy Agency (IRENA): Abu Dhabi, UAE, 2018.
2. Melikoglu, M. Current status and future of ocean energy sources: A global review. *Ocean Eng.* **2018**, *148*, 563–573. [CrossRef]
3. Uihlein, A.; Magagna, D. Wave and tidal current energy–A review of the current state of research beyond technology. *Renew. Sustain. Energy Rev.* **2016**, *58*, 1070–1081. [CrossRef]
4. Heath, T.V. A review of oscillating water columns. *Philos. Trans. R. Soc. A Math. Phys. Eng. Sci.* **2012**, *370*, 235–245. [CrossRef] [PubMed]
5. Torre-Enciso, Y.; Ortubia, I.; De Aguileta, L.L.; Marqués, J. Mutriku wave power plant: From the thinking out to the reality. In Proceedings of the 8th European Wave and Tidal Energy Conference, Uppsala, Sweden, 7–10 September 2009; Volume 710, pp. 319–329.
6. Falcao, A.F.; Henriques, J.C. Oscillating-water-column wave energy converters and air turbines: A review. *Renew. Energy* **2015**, *85*, 1391–1424. [CrossRef]
7. Liang, X. Emerging power quality challenges due to integration of renewable energy sources. *IEEE Trans. Ind. Appl.* **2016**, *53*, 855–866. [CrossRef]
8. Blavette, A.; O'Sullivan, D.L.; Alcorn, R.; Lewis, T.W.; Egan, M.G. Impact of a medium-size wave farm on grids of different strength levels. *IEEE Trans. Power Syst.* **2014**, *29*, 917–923. [CrossRef]
9. Blavette, A.; Kovaltchouk, T.; Rongère, F.; de Thieulloy, M.J.; Leahy, P.; Multon, B.; Ahmed, H.B. Influence of the wave dispersion phenomenon on the flicker generated by a wave farm. In Proceedings of the 12th European Wave and Tidal Energy Conference, Cork, Ireland, 27 August–1 September 2017.
10. Rajapakse, G.; Jayasinghe, S.; Fleming, A.; Negnevitsky, M. Grid integration and power smoothing of an oscillating water column wave energy converter. *Energies* **2018**, *11*, 1871. [CrossRef]
11. Henriques, J.C.C.; Sheng, W.; Falcão, A.F.O.; Gato, L.M.C. A comparison of biradial and wells air turbines on the Mutriku breakwater OWC wave power plant. In Proceedings of the ASME 36th International Conference on Ocean, Offshore and Arctic Engineering, Trondheim, Norway, 25–30 June 2017. [CrossRef]
12. Faÿ, F.X.; Kelly, J.; Henriques, J.; Pujana, A.; Abusara, M.; Mueller, M.; Touzon, I.; Ruiz-Minguela, P. Numerical Simulation of Control Strategies at Mutriku Wave Power Plant. In Proceedings of the ASME 37th International Conference on Ocean, Offshore and Arctic Engineering, Madrid, Spain, 17–22 June 2018. [CrossRef]

13. Rea, J.; Kelly, J.; Alcorn, R.; O'Sullivan, D. Development and Operation of a Power Take Off Rig for Ocean Energy Research and Testing. In Proceedings of the 9th European Wave and Tidal Energy Conference, Southampton, UK, 5–9 September 2011.
14. Kelly, J.F.; Christie, R. Applying Hardware-in-the-Loop capabilities to an ocean renewable energy device emulator. In Proceedings of the IEEE 12th International Conference on Ecological Vehicles and Renewable Energies (EVER), Monte Carlo, Monaco, 11–13 April 2017; pp. 1–7. [CrossRef]
15. International Electrotechnical Commission. IEC Standard, Publication IEC 61000-4-7, Testing and Measurement Techniques-General Guide on Harmonics and Interharmonics Measurement and Instrumentation for Power Supply Systems and Equipment Connected Thereto. 2009. Available online: https://webstore.iec.ch/publication/4228 (accessed on 20 September 2019).
16. International Electrotechnical Commission. IEC Standard, Publication 61000-2-1, Electromagnetic Environment for Low-Frequency Conducted Disturbances and Signalling in Public Power Supply Systems. 1990. Available online: https://webstore.iec.ch/publication/4127 (accessed on 20 September 2019).

© 2019 by the authors. Licensee MDPI, Basel, Switzerland. This article is an open access article distributed under the terms and conditions of the Creative Commons Attribution (CC BY) license (http://creativecommons.org/licenses/by/4.0/).

Article

Wave-to-Wire Power Maximization Control for All-Electric Wave Energy Converters with Non-Ideal Power Take-Off

Marios Charilaos Sousounis * and Jonathan Shek

School of Engineering, Institute for Energy Systems, The University of Edinburgh, Faraday Building, The King's Buildings, Colin Maclaurin Road, Edinburgh EH9 3DW, UK
* Correspondence: M.Sousounis@ed.ac.uk

Received: 19 May 2019; Accepted: 29 July 2019; Published: 31 July 2019

Abstract: The research presented in this paper investigates novel ways of optimizing all-electric wave energy converters for maximum wave-to-wire efficiency. In addition, a novel velocity-based controller is presented which was designed specifically for wave-to-wire efficiency maximization. In an ideal wave energy converter system, maximum efficiency in power conversion is achieved by maximizing the hydrodynamic efficiency of the floating body. However, in a real system, that involves losses at different stages from wave to grid, and the global wave-to-wire optimum differs from the hydrodynamic one. For that purpose, a full wave-to-wire wave energy converter that uses a direct-drive permanent magnet linear generator was modelled in detail. The modelling aspect included complex hydrodynamic simulations using Edinburgh Wave Systems Simulation Toolbox and the electrical modelling of the generator, controllers, power converters and the power transmission side with grid connection in MATLAB/Simulink. Three reference controllers were developed based on the previous literature: a real damping, a reactive spring damping and a velocity-based controller. All three literature-based controllers were optimized for maximum wave-to-wire efficiency for a specific wave energy resource profile. The results showed the advantage of using reactive power to bring the velocity of the point absorber and the wave excitation force in phase, which was done directly using the velocity-based controller, achieving higher efficiencies. Furthermore, it was demonstrated that maximizing hydrodynamic energy capture may not lead to maximum wave-to-wire efficiency. Finally, the controllers were also tested in random sea states, and their performance was evaluated.

Keywords: wave energy converters; wave-to-wire modelling; point absorber; direct drive; permanent magnet linear generator

---

1. Introduction

Wave energy has the potential to supply significant amounts of clean and renewable energy to the electrical grid. Adding wave energy into the energy mix of an electrical system can potentially enhance reliability of supply to isolated island communities, while also decreasing transmission losses, as the energy is generated locally. According to [1], the technically available resource is estimated, in the worst-case scenario, to be 146 TWh/yr, with an installed capacity of 500 GW. The total installed and planned capacity of wave energy converter (WEC) projects is approximately 1 GW, which is well below the worst-case scenario for the technically available resource. One of the main challenges wave energy exploitation faces, according to [2], is the technology cost, which can vary between 330 € to 630 €/MWh, and is significantly higher compared to the approximately 142 €/MWh for offshore wind. It is, therefore, of crucial importance to reduce the cost of energy. The approach taken in this research paper in order to reduce the cost of wave energy conversion is to increase the production of energy by implementing control methods that aim at maximizing energy at the point that energy is sold.

In addition, the increase in power output of a WEC should not affect the cost of production, should include system constraints for rated values, and should be able to be easily applied to devices without the need for significant additional hardware.

At present, several different WEC designs exist, and these can be categorized based on the way they extract energy from the waves, depth of operation, distance to shoreline and power take-off (PTO) type. In this research paper, the focus is given to modelling single-body point absorber WECs with all-electric PTO. Single-body absorbers are chosen because they are more mature, can be designed more easily to suit a specific sea state, and a large number of demonstration projects have been implemented [3]. The use of an all-electric PTO is crucial for the final goal of this research. By using an all-electric PTO with a direct drive (DD) permanent magnet linear generator (PMLG), the conversion stages between wave energy and electrical energy are minimized, which gives the opportunity for higher conversion efficiencies [4,5]. In addition to PMLG, linear switched reluctance generators can also be used, as presented in [6]. The use of hydraulic PTO systems is quite common in WECs, but they introduce a hydraulic transmission interface between the prime mover and the electrical generator. The hydraulic transmission may operate at constant or variable pressure, using low- and high-pressure accumulators [7,8]. Another common approach is to use similar components to those used in the wind and tidal energy industry, such as high-speed electrical generators, permanent magnet synchronous, or induction with a gearbox. The disadvantage of this method, for wave energy, is that an additional component is required, usually a winch, to convert the linear motion of the waves to rotational motion. The winch cylinder can be attached to a gearbox to multiply the speed to appropriate levels for the high speed generator [9–11]. Figure 1 summarizes the conversion stages for the different cases above and shows the inherent advantage of using an all-electric approach for the efficiency of the WEC. Even though the DD All-Electric approach is preferred in this research, the control strategies developed for power maximization can be applied to all types of PTOs with some modifications.

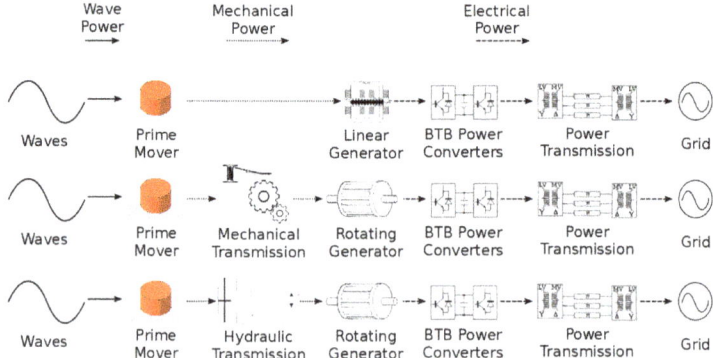

Figure 1. The conversion of power at each stage for different concepts of PTO systems in point absorber WECs.

Many research papers and groups are focused on maximizing the wave energy captured by the WECs. The majority of previous research has been focused on maximizing the capture of wave energy from the prime mover and its conversion to mechanical energy. This research is summarized in [12] for several different designs, and an in-depth explanation is given of the process of maximizing the power captured by the prime mover and states the importance of machine learning and model predictive control (MPC) in wave energy. In [13], the author proposed several causal sub-optimal reactive controllers in order to maximize the power captured from the waves without the need for measuring the frequency of the incoming waves. Similar research presented in [14] investigated an optimum reactive controller for DD WEC with PMLG when the wave frequencies are measured. The authors in [15] tried to eliminate the need for advance knowledge of the wave frequency by implementing several frequency estimation methods. Reactive control for maximizing wave energy capture was also the main focus in [16] for phase control. The authors in [16] also tested their phase controller in an experimental setup by emulating the WEC through a motor. MPC has been the focus of many researchers for optimizing the hydrodynamic efficiency of the WEC. Their use was recently summarized in [17]. However, due to the complexity of the MPC controller, uncertainty over its stability and, most importantly, the high computational needs for their improved performance, researchers have also considered simpler approaches with similar characteristics. The controller proposed in [18] aims at replacing an MPC with filters and transfer functions, which are less computationally intensive to run in real time. The results suggest that the performance of such a controller may be sub-optimal, but with only small differences when compared to an MPC. Another control strategy was presented in [10], and explored the option of linking the power factor of the generator to the load with the aim of highest possible average power with the least peak power generated. In this paper, the reactive power controllers developed in [13] will be presented theoretically and designed for a specific test case.

An aspect of crucial importance in WEC modelling and control is the use of constraints. Several constraints can be set, such as peak phase voltage, peak phase current, peak electrical power output from the generator, maximum velocity and displacement of the translator. The authors in [19] set out some of the electrical requirements in modelling a WEC. They present a novel flux weakening control method for PMGs in order to limit the peak power output to the rated value. The flux weakening method for power limitation, along with constraints set to an MPC for maximum force input, is discussed in [7]. In this paper, a different approach will be taken to limit the power of the generator by applying a mechanical brake. The power limitation constraint is considered in the optimization process and the use of the mechanical brake is minimized. The maximum velocity can monitored by manipulating the PTO force and the maximum displacement is ensured by having an end-stop which provides an increasing opposing force to the translator of the PMLG [13,20]. However, the use of the end-stop for limiting the displacement should only be considered as a back-up, due to the slamming forces that may be applied by the end-stop system to the translator. The controller of the electrical generator should be able to constrain the displacement of the translator, as was presented in [18].

The aim of this paper is to present a comprehensive method for optimizing a wave-to-wire WEC controller in order to maximize the average power delivered to the connection point. For this purpose, existing controllers were tested and optimized but also a novel velocity-based controller was developed. This differs significantly from maximizing the power capture from the waves. This is due to the fact that, in order to maximize hydrodynamic power, the velocity of the translator must be in phase with the excitation force with an appropriate amplitude. In an irregular sea state, this process may consume significant power to move the translator and therefore limiting the average electrical power produced despite the increased hydrodynamic efficiency. Wave-to-wire optimization can also be realized by observing Figure 1. The efficiencies of all the conversion stages between the waves and the grid have to be taken into account in order to achieve maximum electrical output at the grid terminals. This is the point at which the modelling of a non-ideal PTO is crucial. The effect of a non-ideal PTO has been studied in several hydraulic-based WEC systems. In [21], the authors showed that a two-body point absorber WEC can have higher efficiencies compared to a single-body WEC when a non-ideal PTO system is considered. Their approach was based on mathematical modelling and estimated PTO efficiencies for several different PTO damping and stiffness coefficients as well as different body sizes. A similar approach was used in [22] for single-body WEC but the estimated efficiencies assumed a grid connection. In [23], the hydraulic-based WEC system was implemented in a laboratory setup and showed that in all cases an optimally designed reactive controller is beneficial even if the PTO efficiency is as low as 60%. The researchers in [24] thoroughly reviewed wave-to-wire control strategies and stated the importance of having coupled hydrodynamic, mechanical and electrical systems in order to accurately model the operation of WEC systems. The researchers in [7] and [25] developed a high-fidelity wave-to-wire WEC system using a hydraulic PTO, and validated the electrical generator and the control structure. In [26], the MPC is implemented in order to maximize the electrical power output of the generator for a DD system with electrical PTO. The authors defined different stage efficiencies, which helped them perform a cost estimation of the WEC under different cases. Finally, the researchers in [27] presented a novel hill-climbing control method for maximizing the electrical power output.

To achieve the aim of maximizing the power at the point of grid connection, a detailed wave-to-wire single-body point absorber WEC system was developed using multibody physical modelling. Using multibody physical modelling is a novel aspect compared to the above-mentioned literature, because all of the different conversion stages are considered to be non-ideal, including the hydrodynamics, the PTO and the electrical part with the grid connection. The hydrodynamic, mechanical and electrical modelling process is discussed in Section 2. Section 3 presents the controller considered in this research and the subsequent optimization process for power maximization at the grid connection. In addition, Section 3 presents in detail the velocity-based controller developed specifically for the wave-to-wire power maximization, which was based on the research presented in [28]. Section 4 presents the implementation of the methodology of Sections 2 and 3 using a realistic test case based on the waters of China. Finally, Section 5 summarizes the contributions of this research paper.

2. Modelling an All-Electric Wave Energy Converter

The all-electric wave-to-wire WEC model developed for the purposes of this research can be seen in Figure 2. The wave resource model generates a wave excitation force (F_{exc}), which is used as an input to the mechanical model of the point absorber (pa). The hydrodynamic parameters of the point absorber are generated by NEMOH, a Boundary Element Methods code dedicated to the computation of first order wave loads on offshore structures developed by researchers at Ecole Centrale de Nantes in France. The electrical generator model uses the displacement of the point absorber to generate the PTO force (F_{pto}), which is fed back to the mechanical model. The voltage at the generator terminals is controlled by a voltage source converter (VSC) using a generator controller strategy. The generator control strategies will be discussed in detail in Section 3. On the grid side, the inverter is controlled by the Voltage Oriented Controller (VOC). In this study, power transmission components have also been included prior to grid connection.

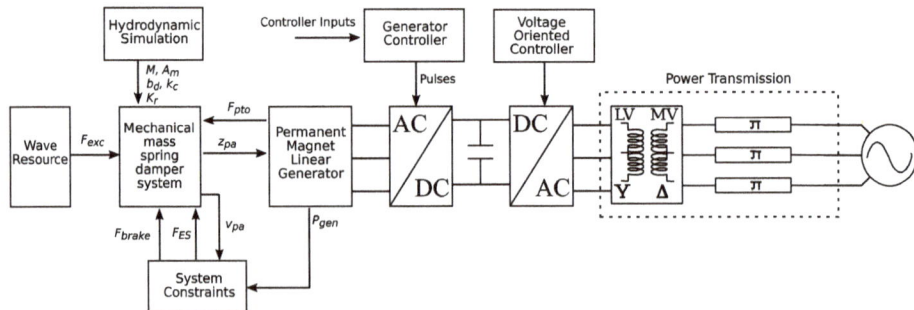

Figure 2. Wave-to-wire WEC model developed primarily in MATLAB/Simulink. The hydrodynamic simulations were performed in NEMOH. The model incorporates a DD All-Electric PTO system, system constraints, power transmission and grid connection.

2.1. Hydrodynamic Simulation

The hydrodynamic simulation of the point absorber was performed by the Edinburgh Wave Systems Simulation Toolbox (EWST) [29], which uses NEMOH and can also generate a refined mesh for simple structures appropriate for hydrodynamic simulation. Figure 3 shows the design and refined mesh of a cylinder generated by EWST which will be used as a point absorber. The radius of the cylinder is 1.05 m and the draft is 3 m. The total height of the cylinder is 4 m.

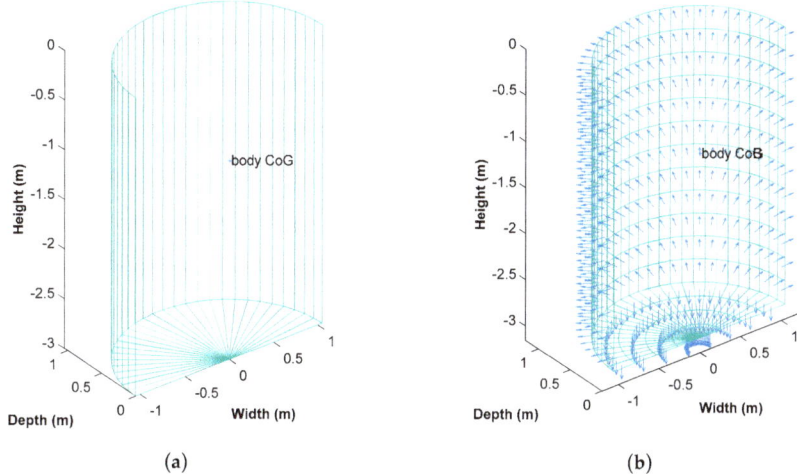

Figure 3. Mesh design for the submerged part of a cylindrical buoy generated by EWST to perform NEMOH hydrodynamic simulation. (**a**) NEMOH mesh of the cylindrical buoy. The center of gravity (CoG) is visible; (**b**) Refined mesh of the cylindrical buoy with the normal pointing out. The center of buoyancy (CoB) is visible.

The hydrodynamic simulation is performed for a wide range of wave frequencies and produces several outputs from which the mass (M), added mass (A_m), linear restoring coefficient or stiffness (k_c), linear damping (b_d) and the radiation force (F_{rad}) are used in the point absorber model to form the mass-spring-damper system. Radiation damping and added mass as a function of the wave frequency input are given in Figure 4. The constant hydrodynamic parameters are given in Table 1. As part of this work, a simple cylinder design was considered as a point absorber. This is due to the fact that the main aim of this research is to demonstrate a control process to maximize wave-to-wire power through the control of the electrical generator. Optimizing the structure of the point absorber was deemed outside the main scope of this study.

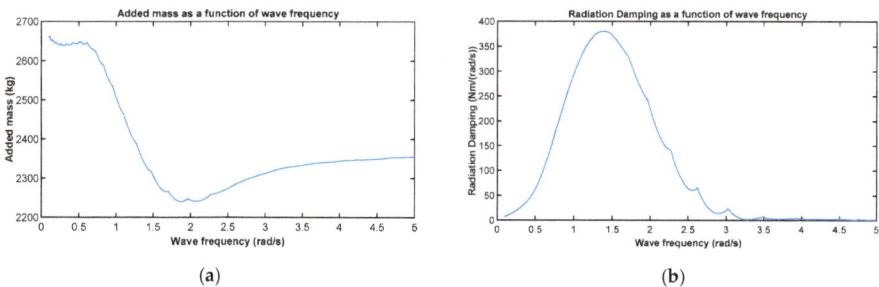

Figure 4. Hydrodynamic simulation outputs as a function of wave frequency for the cylindrical buoy of Figure 3: (**a**) Added mass; (**b**) Radiation damping.

Table 1. Hydrodynamic and point absorber constant parameters for the cylindrical buoy under study.

Symbol	Quantity	Value	Units
M	Mass	10,630	kg
A_m	Added mass at 5 rad/s	2354	kg
b_d	Linear damping	133	Ns/m
k_c	Stiffness	34,760	N/m
w_n	Natural frequency	1.8083	rad/s
pa_r	Buoy radius	1.05	m
pa_h	Buoy height	4	m
F_{rated}	Rated mechanical force	30,000	N
b_{ES}	End-stop damping	45	Ns/m
k_{ES}	End-stop stiffness	11,650	N/m
pa_{zmax}	Translator maximum displacement	1.5	m

2.2. Point Absorber Model and System Constraints

The single-body point absorber model in this paper is represented using a mass-spring-damper system which moves on the heave (z-axis) direction only. It is assumed that the salt water is inviscid and incompressible, and the flow is irrotational. Based on the above assumptions, the interaction between the point absorber model and the incoming waves can be studied using linear wave theory. The time-domain equation of motion of the point absorber is given in Equation (1), and the mechanical system developed in MATLAB/Simulink is given in Figure 5a.

$$Ma_{pa}(t) = F_{exc}(t) - F_{pto}(t) - F_{rad}(t) - z_{pa}(t)k_c - v_{pa}(t)b_d - F_{brake} - F_{ES} \quad (1)$$

where a_{pa} is the acceleration and v_{pa} the velocity of the point absorber. F_{rad} is the force as a result of wave radiation and is estimated using a State Space model. The parameters of the State Space model are taken from the hydrodynamic simulation and are based on the convolution integral formulation of the radiation impulse response function (K_r) of the point absorber and is shown in Equation (2).

$$F_{rad}(t) = \int_0^t K_r(t-\tau)v_{pa}(\tau)d\tau \quad (2)$$

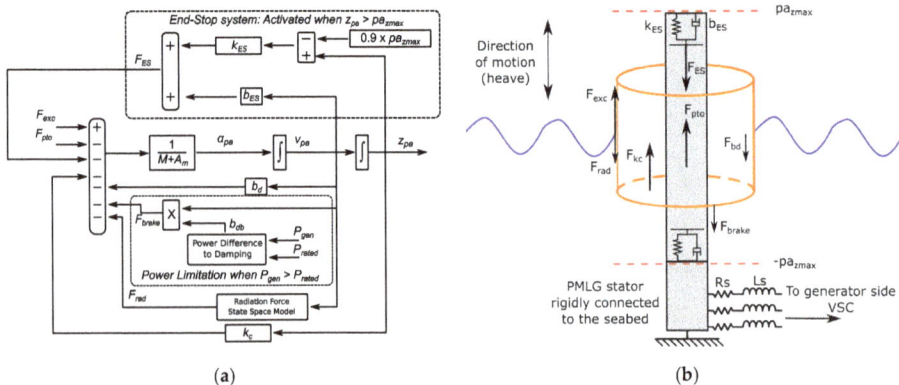

Figure 5. (a) Block diagram of the mechanical mass-spring-damper model of the single buoy point absorber described in (1) developed in MATLAB/Simulink; (b) Diagram of the WEC and PTO configuration. All forces are acting on the heaving buoy (one degree of freedom).

The added mass (A_m), linear damping (b_d) and stiffness (k_c) are considered constants and their values are given in Table 1, along with point absorber parameters. Linear damping generates a linear damping force (F_{bd}) and stiffness generates a buoyancy force (F_{kc}). The F_{exc} is generated by the wave resource model, and the F_{pto} is generated by the PMLG, which is discussed in detail in Section 2.3.

Two systems that constrain the operation of the point absorber system have been implemented. The first constraint is the end-stop (subscript ES) system, which limits the displacement (z_{pa}) to its maximum value (pa_{zmax}). The ES system operates only when the z_{pa} has reached 90% of the pa_{zmax}, and generates the end-stop force (F_{ES}), which slows down the translator. The F_{ES} is generated by taking into account that the end-stop spring has a damping coefficient (b_{ES}) and a stiffness coefficient (k_{ES}). The values for the end-stop system are given in Table 1. The second constraint implemented is the electrical power limitation mechanism. In order to avoid overrated devices, keep the cost of the device low, and avoid excessive peaks of current that could cause damage to the WEC, a power limitation mechanism is essential. In this research paper, the power limitation mechanism is based on a simple braking force (F_{brake}) that is generated when the instantaneous electrical power generated from the PMLG (P_{gen}) is larger than the rated power of the device (P_{rated}). The power difference is converted to a damping coefficient (b_{db}), which can be implemented in reality by using standard brakes. The conversion of the power difference to b_{db} can be modelled either with a linear relationship or with a proportional-integral (PI) controller. For this study, using a power limitation mechanism is of crucial importance. To maximize the electrical power delivered to the grid, it should to be taken into account that sudden peaks in wave power may have to be limited, and therefore will not result in an increase in electrical generation. It has to be noted that even though the optimized controllers developed in this research paper take into account the power limitation for power production the same does not apply for the displacement constraint. Further development of the system will focus on making the controllers aware of the displacement constraint and avoid the end-stop system slamming forces. The PMLG and the parameters of the electrical generator are discussed in Section 2.3 and shown in Table 2, respectively.

Table 2. Permanent magnet linear generator parameters.

Symbol	Quantity	Value	Units
P_{rated}	PMLG rated power	30,000	W
N_t	Number of turns around a tooth	250	-
τ_p	Pole pitch	0.1	m
Φ	Flux in the tooth	0.1073	Wb
Rs	Stator resistance per phase	2.9667	Ω
Ls	Stator inductance per phase	0.0789	H

2.3. The Permanent Magnet Linear Generator

The PMLG model receives as input the displacement from the point absorber model and produces the F_{pto}, which is used in the mechanical mass-spring-damper system. The PMLG is modelled using Equations (3)–(5) [28,30]:

$$\begin{cases} EMF_a = -N_t \frac{\pi}{\tau_p} \Phi \sin\left(\frac{\pi}{\tau_p} z\right) \frac{dz}{dt} \\ EMF_b = -N_t \frac{\pi}{\tau_p} \Phi \sin\left(\frac{\pi}{\tau_p} z - \frac{2\pi}{3}\right) \frac{dz}{dt} \\ EMF_c = -N_t \frac{\pi}{\tau_p} \Phi \sin\left(\frac{\pi}{\tau_p} z - \frac{4\pi}{3}\right) \frac{dz}{dt} \end{cases} \quad (3)$$

$$F_{pto} = \frac{3}{2} \frac{\pi}{\tau_p} \Phi N_t I \cos \varphi \quad (4)$$

$$\begin{cases} i_a = -I\sin\left(\frac{\pi}{\tau_p}z + \varphi\right) \\ i_b = -I\sin\left(\frac{\pi}{\tau_p}z - \frac{2\pi}{3} + \varphi\right) \\ i_c = -I\sin\left(\frac{\pi}{\tau_p}z - \frac{4\pi}{3} + \varphi\right) \end{cases} \quad (5)$$

The EMF of the generator is produced by implementing (3). N_t is the number of turns around a tooth, Φ is the flux in the tooth and τ_p is the pole pitch. As shown in (3), the EMF of the PMLG is a function of the displacement of the translator, which is taken from the point absorber model discussed in Section 2.2. The EMF is transferred to the SimPowerSystems physical modelling environment, and the voltage at the terminals of the PMLG is acquired by taking into account the stator resistance (R_s) and inductance (L_s). The F_{pto} can be calculated using Equation (4), where I is the peak current and the current leads the EMF with an angle φ. The generation of the three-phase currents is given in Equation (5) and is a function of displacement.

The voltage at the PMLG terminals is controlled through the generator-side VSC, which operates based on a control strategy. Several control strategies have been discussed in this paper in Section 1. The main aim of this paper is to study and explore generator-side control strategies that will aim to maximize electrical power at the point of grid connection. This process is shown in detail in Section 3. In Section 2.4, the modelling and design of the grid-side power converter controller (Voltage Oriented Controller) is presented, and this ensures power transfer to the grid side. Finally, Section 2.4 also gives the details of the power transmission system for a single WEC.

2.4. Voltage Oriented Controller and Power Transmission

The power generated by the WEC is delivered to the power transmission system using Back-to-Back (BTB) power converters. The grid-side VSC of the BTB power converters is connected to the grid through several components; a line reactor to reduce line current distortion, a filter that reduces harmonics and a step-up transformer from 400 V to 11 kV. The inverter is controlled by a PWM scheme called voltage oriented control (VOC) with decoupled controllers, which maintains a constant DC link voltage, constant frequency output of 50 Hz on the AC side and control over the amount of reactive power flowing based on grid requirements. The switching frequency of the PWM scheme is set at 2.5 kHz. The VOC scheme is based on the time-domain equations given in (6) [31].

$$\begin{cases} V_{di}(t) = -\left[K_P^I\left(i_{dg}^* - i_{dg}\right) + K_I^I\int\left(i_{dg}^* - i_{dg}\right)dt\right] + \omega_g L_g i_{qg} + V_{dg} \\ V_{qi}(t) = -\left[K_P^I\left(i_{qg}^* - i_{qg}\right) + K_I^I\int\left(i_{qg}^* - i_{qg}\right)dt\right] - \omega_g L_g i_{dg} + V_{qg} \end{cases} \quad (6)$$

where ω_g is the angular frequency of the grid, L_g is the line inductance between the VSC and the point of measurement, i_{dg} and i_{qg} are the dq-axis currents at the point of measurement, i_{dg}^* and i_{qg}^* are the dq-axis reference currents, V_{dg} and V_{qg} are the dq-axis voltages at the point of measurement. K_P^I and K_I^I are the "Decoupled Control" proportional and integral gains of the PI controller. V_{di} and V_{qi} are the reference voltages in the dq-axis frame which control the grid-side VSC. V_{di} and V_{qi} are converted to three-phase reference voltages $V_{abc}*$ with the inverse park transformation by using the angle at the point of measurement θ_g. Figure 6 depicts the block diagram of the VOC scheme as it was implemented in MATLAB/Simulink.

The different parts of the VOC scheme can be seen in Figure 6. Three distinct sub-blocks can be identified in the VOC scheme. The "Reactive power control" takes as input the potential reactive power requirements from the grid side in order to produce the appropriate q-axis reference current. In the VOC scheme, the q-axis current controls the flow of reactive power to and from the VSC and d-axis current controls the flow of active power. Therefore, by changing the i_{qg}^*, the reactive power flow can be controlled directly. For the purposes of this research, it is assumed that there is no need for the VSC to generate any reactive power, and therefore $i_{qg}^* = 0$. The "DC Voltage control" ensures that the DC link voltage is kept constant at 800 V. To do that, the PI controller in the "DC Voltage control" has to produce an appropriate i_{dg}^*, which will ensure that the active power flowing from the generator side will be transferred to the grid side. Finally, the "Decoupled control" compares reference and actually measured currents to produce the reference voltages. The parameters of the BTB power converters and VOC scheme are given in Table 3.

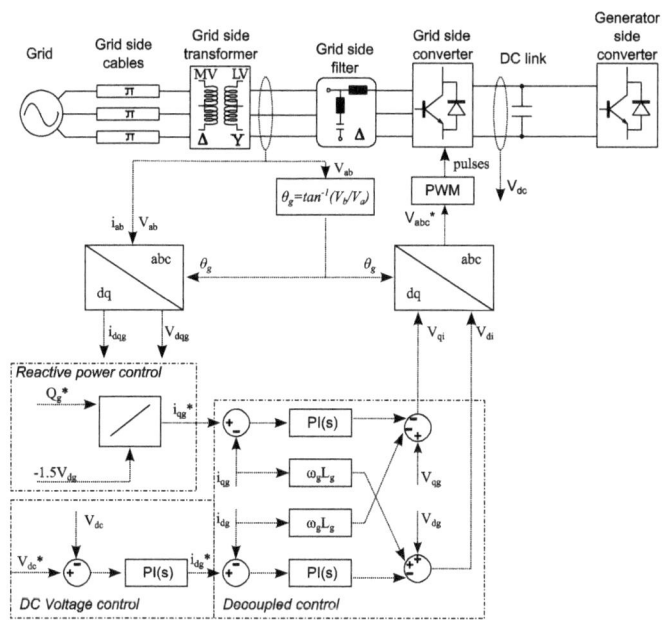

Figure 6. Block diagram of the Voltage Oriented Controller as it was implemented in MATLAB/Simulink. The point of measurement for the electrical quantities is at the low voltage side of the power transmission transformer after the grid-side filter.

Table 3. Parameters for the BTB power converters and VOC.

Symbol	Quantity	Value	Units
f_g	Grid side frequency	50	Hz
ω_g	Grid side angular frequency	314	rad/s
V_{dc}^*	DC link reference voltage	800	V
Q_g^*	Reactive power reference	0	var
L_g	Line inductance	0.0019	H
R_{on}	IGBT on-state resistance	0.01493	Ω
V_{ce}	IGBT collector emitter saturation voltage	1.9	V

The power transmission system is composed of a step-up transformer to increase the voltage from 400 V to 11 kV, as well as three-phase cables that are directly connected to an 11 kV strong grid. The three-phase cables are modelled as a network of π-sections. One section per km of cable length was chosen in order to represent the harmonic resonance. The parameters of the power transmission system are given in Table 4.

Table 4. Parameters of the power transmission system.

Symbol	Quantity	Value	Units
P_{gt}	Transformer nominal power	40,000	W
V_{ygt}	Wye line RMS nominal voltage	400	V
V_{Dgt}	Delta line RMS nominal voltage	11,000	V
R_{ygt}	Wye resistance	0.0058	pu
L_{ygt}	Wye inductance	0.0612	pu
R_{Dgt}	Delta resistance	0.0038	pu
L_{Dgt}	Delta inductance	0.0151	pu
l_C	Cable length	3	km
R_C	Cable resistance per km	0.223	Ω/km
L_C	Cable inductance per km	0.43	mH/km
C_C	Cable capacitance per km	0.24	µF/km

3. Control Strategies for Wave-to-Wire Power Maximization

The main aim of this research paper is to present ways of maximizing the power delivered to the grid through appropriate generator-side control. As shown in Section 1, a lot of focus has been given to maximizing the hydrodynamic power captured by a WEC. However, this is not always the optimum solution for power generation, since reactive power is needed when maximizing hydrodynamic power. This means that the generator-side VSC needs to supply the generator with power, driving it as a motor, in order to bring the velocity of the prime mover in phase with the wave excitation force. Another important aspect for an optimum WEC controller is the availability of measurements and, more specifically, the incoming wave frequency measurement. Due to the difficulty of measuring the frequency of the incoming waves, the need for additional hardware and increased cost, this paper focusses on implementation of control strategies that do not require knowledge of the incoming wave frequency. The generator-side controller for WECs with PMLGs can be separated into two parts:

- The reference F_{pto} controller; and
- The pulse generator for the VSC.

The first part produces an instantaneous reference PTO force ($F_{pto}*$), which forces the point absorber model to move based on the requirements of the control strategy. Depending on the control strategy, $F_{pto}*$ can bring, for example, the point absorber velocity into phase with the excitation force from the waves to maximize hydrodynamic power, or can only control the damping to maximize the power transfer without any phase control. As mentioned in Section 2, the W2W model is based on physical modelling of the mechanical and electrical quantities. This means that the $F_{pto}*$ generated by the controller is not directly applied to the point absorber model which is a common practice. The $F_{pto}*$ from the control strategy is used as an input to generate pulses for the generator-side VSC. The pulses that control the generator-side VSC will force the PMLG to produce the appropriate PTO force. The block diagram of the generator-side control strategy and the pulse generator are depicted in Figure 7. The pulse generator for the generator-side VSC is based on the zero d-axis current (ZDC) controller for PMLG.

The ZDC controller requires as inputs $F_{pto}*$, z_{pa}, the reference d-axis current ($i_{ds}*$), and the measured three-phase current at the terminals of the PMLG (i_{abc}). As shown in Figure 7b, the displacement is converted to electrical angle, and the $F_{pto}*$ is converted to reference q-axis current ($i_{qs}*$). Reference currents, $i_{ds}*$ and $i_{qs}*$, are compared to actually measured currents to produce errors. The current errors are fed to PI controllers to generate the appropriate reference voltage signals V_q* and V_d* to minimize the current error. The reference voltages are used as input to a PWM scheme for pulse generation for the generator-side VSC.

Regarding Figure 7a, the point absorber constant properties and instantaneous quantities depend on the specific reference PTO force control strategy and includes the properties given in Table 1 and instantaneous values of measured displacement, velocity and acceleration of the point absorber. In this section, three different types of reference F_{pto} controllers will be presented theoretically and tested for their ability to deliver maximum power to the grid. These control strategies are the real damping, reactive spring-damping and velocity controller. For all three different reference F_{pto} controllers the ZDC controller remains the same as a pulse generator for the generator-side VSC.

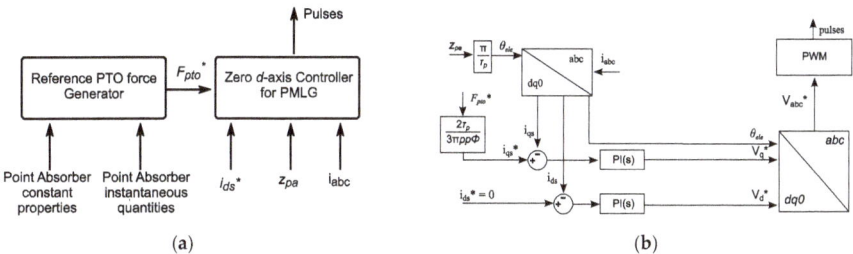

Figure 7. Block diagrams of (**a**) generator-side control strategy and (**b**) zero d-axis controller.

3.1. Real Damping Controller

The Real Damping Controller (RdC) is one of the most common reaction force control methods. The RdC only provides damping PTO force, which affects the amplitude of the velocity, but not the phase of the velocity. By providing damping PTO force, only real power is produced. The main advantage of this control method is that it is simple and easy to implement. In this research paper, it is implemented based on [13] by assuming that the PTO impedance only has a real part. The impedance matching ensures maximum power transfer from the waves to the mechanical mass-spring damper system. The reference PTO force generator for the RdC can be described using Equation (7).

$$F_{pto}^{RdC} = v_{pa} \times \sqrt{b^2(\omega_p) + (\omega_p[M + A_m(\omega_p)] - k_c/\omega_p)^2} \tag{7}$$

where F_{pto}^{RdC} is the reference PTO force from the RdC method, $b(\omega_p)$ is the total damping (radiation and linear) at the wave frequency peak energy in a specific wave climate. Using a variable wave frequency ω_p to calculate the damping and added mass coefficients would lead to improved wave energy capture, but this would require continuous frequency measurements of the incoming waves. Using a single frequency value at which most energy exists for a specific wave climate will allow most of the wave energy for this specific climate to be efficiently converted to mechanical energy, but less energy will be converted overall compared to the use of a variable wave frequency. Simulation results of the W2W WEC system developed using the RdC method are given in Figure 8.

Figure 8a shows the wave elevation generated by a JONSWAP spectrum with significant wave height (H_s) 1.45 m and peak energy period (T_e) 6 s. The wave power density for this spectrum in deep water can be calculated using Equation (8) in kW/m.

$$P_W = 0.5 \times H_s^2 \times T_e \tag{8}$$

Using Equation (8), the incoming wave power can be calculated for the buoy under study in this specific sea state. Figure 8b shows the reference PTO force generated by the RdC method (F_{pto}^{RdC}) and the actually generated PTO force from the linear generator. It can be observed that the F_{pto} closely follows the F_{pto}^{RdC}, but they are not identical. Some differences appear when the force is above 12 kN. This may be the result of a sudden wave elevation peak. Figure 8c shows the synchronization of the velocity of the point absorber and wave excitation force. The RdC method does not provide any phase control, and therefore the synchronization is based only on the point absorber's natural frequency and the incoming wave frequencies. The point absorber's natural frequency is given in Table 1 as 1.8083 rad/s, which is larger than the 1.0467 rad/s peak energy frequency of the sea state used. For the calculation of the F_{pto}^{RdC} using Equation (7), ω_p was equal to 1.0467 rad/s, which is the peak energy frequency of the sea state used. Figure 8d presents the power output at different stages of the system. The peak instantaneous electrical generator power output (P_{gen}) is 19 kW, whereas the average values for the hydrodynamic power (P_{hydro}^{avg}), generator power (P_{gen}^{avg}) and grid power (P_{grid}^{avg}) are 1.9147 kW, 1.6909 kW and 1.4701 kW, respectively. This leads to a PMLG efficiency (η_{PMLG}) of 88.3% and electrical system efficiency (generator output to grid) of 86.9% (η_{Ele}). If we assume that the power of the incoming waves is P_W, then the hydrodynamic efficiency (η_{hydro}) can also be calculated as being 10.1%. The W2W efficiency (η_{W2W}) is 7.8%. The different efficiencies can be calculated using Equation (9).

$$\begin{cases} \eta_{hydro} = \dfrac{P_{hydro}^{avg}}{P_W} \times 100 \\ \eta_{PMLG} = \dfrac{P_{gen}^{avg}}{P_{hydro}^{avg}} \times 100 \\ \eta_{Ele} = \dfrac{P_{grid}^{avg}}{P_{gen}^{avg}} \times 100 \\ \eta_{W2W} = \dfrac{P_{grid}^{avg}}{P_W} \times 100 \end{cases} \qquad (9)$$

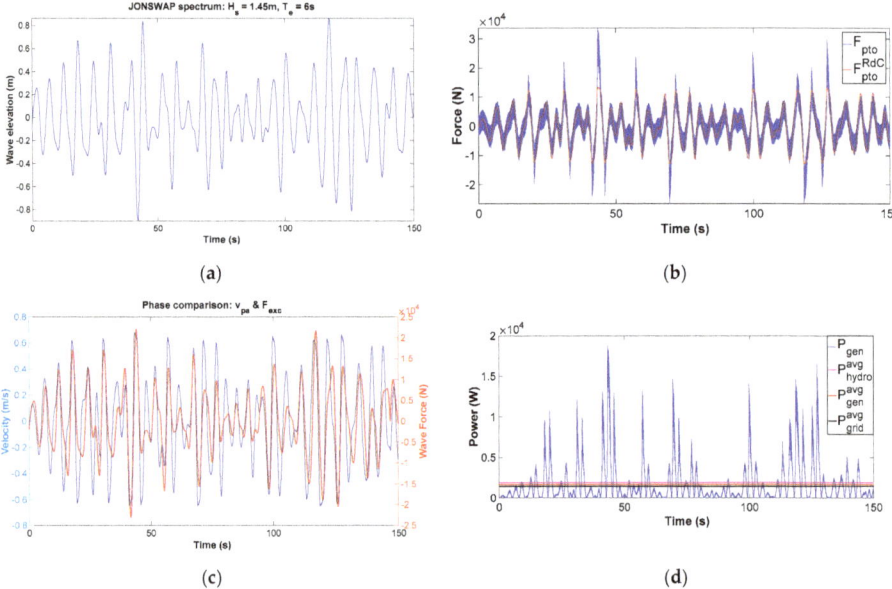

Figure 8. Simulation results using the RdC method: (**a**) Wave resource used as input; (**b**) Reference and actual PTO force; (**c**) Phase comparison between the velocity and wave excitation force; (**d**) Instantaneous generator power and average hydrodynamic, generator and grid power.

Power Maximization Process for the Real Damping Controller

The main idea behind the power maximization process for the RdC is that Equation (7) has several constant parameters that are based on the point absorber constant hydrodynamic properties. Therefore, the constant hydrodynamic properties are also used for the calculation of the reference PTO force and this is to achieve maximum hydrodynamic efficiency η_{hydro}.

However, several uncertainties in the constant hydrodynamic parameters may lead to sub-optimal performance. For example, the calculation of the total damping may not be accurate, since it has to contain hydrodynamic damping, which is not calculated as a single value, but rather as a state space model, and the linear damping of the PMLG. Difficulty in calculating these values will certainly lead to reduced performance. Another factor in the power maximization process is that, in order to reduce the cost of energy of WEC devices, it is necessary to maximize the power exported, rather than the hydrodynamic power. It is often assumed that by maximizing hydrodynamic power, the whole system power is maximized, and constant efficiencies for the electrical system are assumed. This is not always the case, as several parameters can affect the electrical power output and electrical efficiencies. For example, a sudden increase in the output current may lead to increased electrical losses due to the rise of temperature in the electrical equipment. Electrical equipment that is sensitive to increases in temperature due to the increased current includes both the power converters and the power transmission cables. Furthermore, as was previously stated in this study, the electrical power constraint is also included to avoid power output that can damage the electrical equipment. Finally, the location each WEC is installed has a specific wave energy climate. This wave energy climate can be measured and quantified so that the WEC installed at this specific wave climate may perform adequately. To summarize, the power maximization process is needed because:

- The constant hydrodynamic parameters in the calculation of the reference PTO force may not be accurate.
- The objective is to increase the power exported to the grid and not necessarily the power captured by the waves. The behavior of the electrical generator can change the losses in the electrical system.
- The mechanical and electrical constrains have to be included in the generation of the reference PTO force.
- Each WEC is installed for a specific wave climate. The reference PTO force has to be modified to reflect power maximization for this specific wave climate.

To perform power maximization, Equation (7) is re-written in Equation (10). Two constant hydrodynamic parameters are identified as the ones that are difficult to calculate and affect the calculation of the F_{pto}^{RdC} significantly, the damping and the stiffness.

$$F_{pto}^{RdC} = v_{pa} \times \sqrt{b^{RdC2} + \left(\omega_p[M + A_m(\omega_p)] - k_c^{RdC}/\omega_p\right)^2} \qquad (10)$$

In Equation (10), the damping b^{RdC} and stiffness k_c^{RdC} are considered unknowns, and are modified in order to maximize power exported to the grid. Using the constant hydrodynamic parameters as initial guesses, several optimization algorithms can be implemented in order to maximize P_{grid}^{avg}. The optimization algorithms used in this study were implemented in MATLAB and are based on grid search and constrained non-linear algorithms such as sequential quadratic programming. The theory and formulation of these algorithms is outside the scope of this paper. Figure 9 presents the results of the grid search optimization process, which focuses on changing the b^{RdC} and k_c^{RdC} with the aim of maximizing power at different points of the W2W WEC system.

It can be observed that k_c^{RdC} affects the output of the WEC system, whereas the effect of b^{RdC} is quite small for the range of values used in the grid search optimization. The maximum power for all cases appears to be around 14,000 N/m, with the damping having minimum impact at this scale. Apart from the grid search optimization algorithm, which is time consuming and is based on fixed steps for the design variables, a gradient descent active-set algorithm was used to calculate the b^{RdC} and k_c^{RdC} for maximum power at the grid. The summary of results for all the algorithms and the reference case are presented in Table 5. It should be noted that the optimization is based on a 150 s simulation using the wave energy resource presented in Figure 8a. Based on the results presented for optimizing the Real Damping Controller, higher W2W efficiencies can be achieved with minimal impact on the peak-to-average ratio of the generator power (P_{gen}^{ratio}). It should be noted that, due to the fact that the WEC system presented has a power limitation mechanism, this is included in the optimization process to achieve maximum power at the point of grid connection.

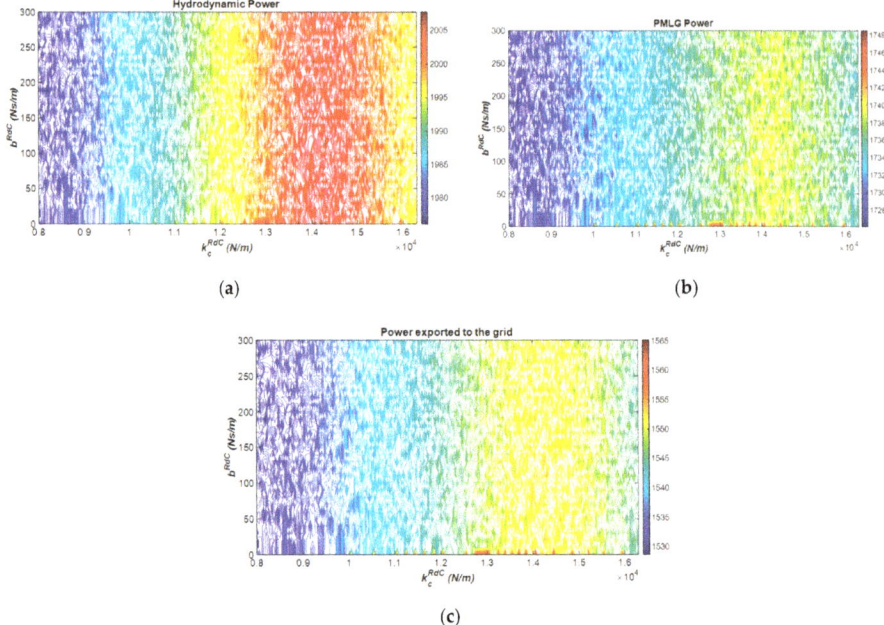

Figure 9. Grid search optimization of b^{RdC} and k_c^{RdC} to maximize (**a**) hydrodynamic power P_{hydro}^{avg}; (**b**) PMLG power P_{gen}^{avg}; and (**c**) power exported to the grid P_{grid}^{avg}.

Table 5. Summarized Real Damping Controller results.

Method	b^{RdC} (Ns/m)	k_c^{RdC} (N/m)	P_{hydro}^{avg} (kW)	P_{gen}^{avg} (kW)	P_{gen}^{ratio}	P_{grid}^{avg} (kW)	η_{W2W} (%)
Reference	274	34,756	1.915	1.691	11.05	1.470	7.77
Grid Search	0	14,050	2.007	1.747	11.78	1.565	8.26
Gradient Descent	0.07	14,144	2.009	1.748	11.76	1.566	8.28

3.2. Reactive Spring Damping Controller

The Reactive Spring Damping Controller (RsdC) is a sub-optimal complex conjugate controller that uses the constant peak energy frequency ω_p of the sea state and does not include an inertia term in the calculation of the reference PTO force F_{pto}^{RsdC}. The reference PTO force calculation for the RsdC is described in Equation (11).

$$F_{pto}^{RsdC} = v_{pa} \times b(\omega_p) + z_{pa} \times \left\{ \omega_p^2 \left[M + A_m(\omega_p) \right] - k_c \right\} \quad (11)$$

Simulation results of the W2W WEC system developed using the RsdC method are given in Figure 10.

The wave energy resource shown in Figure 10a is the same as the one for the Real Damping Controller. In Figure 10b, F_{pto}^{RsdC} is compared with the actual F_{pto} generated by the PMLG. The PMLG PTO force does not follow the reference signal in all the cases. Several factors can affect this performance including PMLG design parameters such as the phase inductance. An additional observation is that F_{pto} has much higher values compared to Figure 8b. Regarding phase synchronization between the wave excitation force and the velocity of the point absorber, it is evident that the RsdC affects the v_{pa} phase, especially when compared to Figure 8c. Apart from the phase control, the amplitude control of the RsdC is significant as well. The velocity of the point absorber reaches 0.6 m/s even at low excitation force amplitudes. Finally, the power at different stages can be seen in Figure 10d. The peak instantaneous electrical generator power output (P_{gen}) is 23.8 kW, whereas the average values for the hydrodynamic power (P_{hydro}^{avg}), generator power (P_{gen}^{avg}) and grid power (P_{grid}^{avg}) are 2.6549 kW, 2.2537 kW and 2.0247 kW, respectively. This leads to η_{PMLG} = 84.9%, η_{Ele} = 89.8%, η_{hydro} = 14.1% and η_{W2G} = 10.7%. The different efficiencies are calculated using Equation (9). In addition, reactive power flow can be seen in Figure 10d. The peak negative power at the generator terminals is around 2 kW, and the average negative power in a 150 s simulation is 680 W. By implementing the RsdC, total power to the grid is increased by 38% compared to RdC, despite the negative power consumed by the generator to perform phase control. In the following section, the RsdC is optimized in a similar way to the RdC in order to maximize the power exported to the grid.

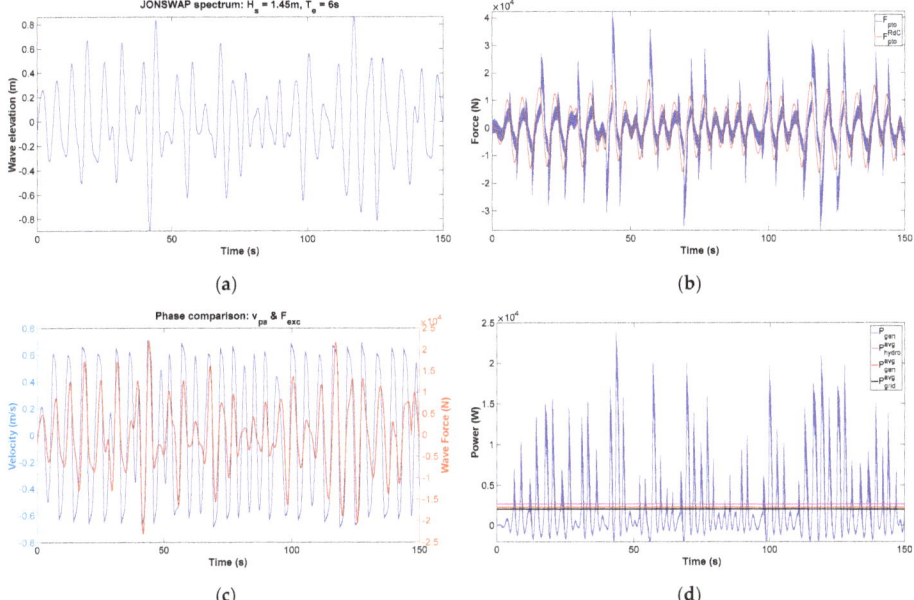

Figure 10. Simulation results using the RsdC method. (a) Wave resource used as input; (b) Reference and actual PTO force; (c) Phase comparison between the velocity and wave excitation force; (d) Instantaneous generator power and average hydrodynamic, generator and grid power.

Power Maximization Process for the Reactive Spring Damping Controller

The main idea behind the power maximization process for the RsdC is that Equation (11) has several constant parameters that are based on the point absorber constant hydrodynamic properties. Therefore, the constant hydrodynamic properties are also used for the calculation of the reference PTO force, and this is done in order to achieve maximum hydrodynamic efficiency, η_{hydro}. The WEC controllers are optimized in order to maximize P_{grid}^{avg}. An additional characteristic of the RsdC compared to the RdC is that it has reactive power, allowing it to carry out phase control of the point absorber. Reactive power control leads to power being consumed by the PMLG in order to bring the velocity of the point absorber into phase with the wave excitation force. The amount of power consumed can significantly affect the total power generated. The reasoning behind performing optimization in the RsdC is similar to the RdC, and is reproduced here with the addition of the control of the reactive power to achieve the desired target:

- The constant hydrodynamic parameters in the calculation of the reference PTO force may not be accurate.
- The objective is to increase the power exported to the grid, and not necessarily the power captured by the waves. The behavior of the electrical generator can change the losses in the electrical system.
- The mechanical and electrical constraints have to be included in the generation of the reference PTO force.
- Each WEC is installed in a specific wave climate. The reference PTO force has to be modified to reflect power maximization for this specific wave climate.
- Reactive power control can bring the point absorber into phase with the wave excitation force for maximum power extraction from the waves, but may lead to excessive power being consumed by the WEC. The amount of reactive power has to be controlled through optimization.

To perform power maximization, Equation (11) is rewritten into Equation (12). Two constant hydrodynamic parameters are identified as being difficult to calculate and as significantly affecting the calculation of F_{pto}^{RsdC}: damping and stiffness.

$$F_{pto}^{RsdC} = v_{pa} \times b^{RsdC} + z_{pa} \times \{\omega_p^2[M + A_m(\omega_p)] - k_c^{RsdC}\} \qquad (12)$$

In Equation (12), damping b^{RsdC} and stiffness k_c^{RsdC} are considered unknowns and are modified in order to maximize power exported to the grid. Using the constant hydrodynamic parameters as initial guesses, several optimization algorithms can be implemented in order to maximize P_{grid}^{avg}. The optimization algorithms used in this study were implemented in MATLAB and are based on grid search and constrained non-linear algorithms such as the sequential quadratic programming. Figure 11 presents the results of the grid search optimization process which focuses on changing the b^{RsdC} and k_c^{RsdC} with the aim of maximizing power at different points of the W2W WEC system.

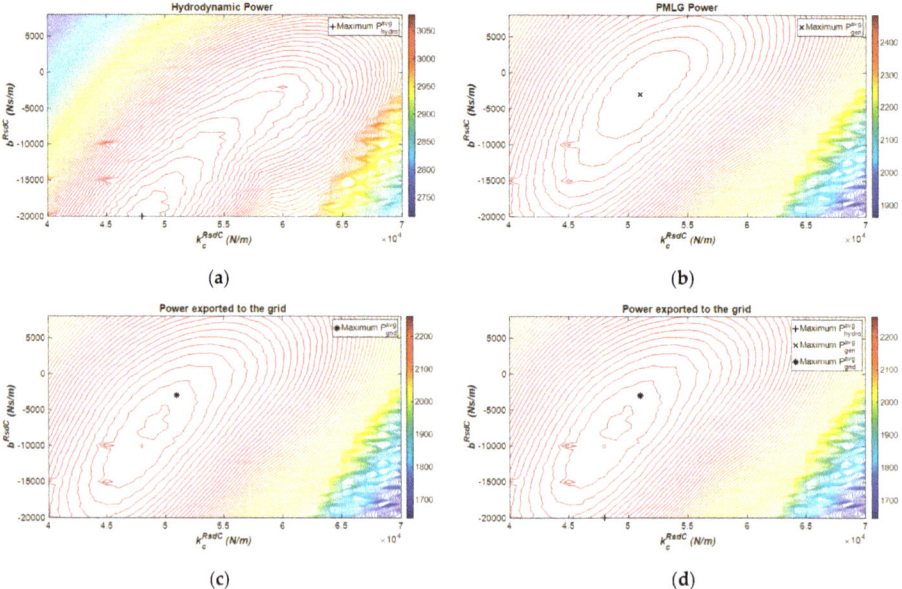

Figure 11. Grid search optimization of b^{RsdC} and k_c^{RsdC} to maximize: (**a**) Hydrodynamic power P_{hydro}^{avg}; (**b**) PMLG power P_{gen}^{avg}; (**c**) Power exported to the grid P_{grid}^{avg}; (**d**) Power exported to the grid with overlapped peak power points of hydrodynamic and PMLG output.

Figure 11 depicts power output of the point absorber (Figure 11a), generator (Figure 11b) and grid (Figure 11c) for different parameters of b^{RsdC} and k_c^{RsdC}. For each case, the b^{RsdC} and k_c^{RsdC} that gives the maximum average power is identified, cross (+) for maximum P_{hydro}^{avg}, cross mark (x) for maximum P_{gen}^{avg}, and asterisk (*) for maximum P_{grid}^{avg}. In Figure 11d, the maximum power points are overlapped on the contour plot of power exported to the grid. It is shown that maximum power is achieved at different combinations of b^{RsdC} and k_c^{RsdC}, which shows how much reactive power can affect the power output at different stages of the system. The maximum P_{gen}^{avg} and P_{grid}^{avg} are at the same b^{RsdC} and k_c^{RsdC} values, which indicates that optimizing generator power output may suffice for W2W optimization, but this is a single case and cannot be generalized.

Apart from the grid search algorithm visualized in Figure 11, a gradient descent algorithm was also used in a similar way as in the RdC. The results for a 150 s simulation are presented in Table 6. Based on the results presented, optimizing the Reactive Spring Damping Controller, higher W2W efficiencies can be achieved with small impact on the P_{gen}^{ratio}.

Table 6. Summarized Reactive Spring Damping Controller results.

Method	b^{RsdC} (Ns/m)	k_c^{RsdC} (N/m)	P_{hydro}^{avg} (kW)	P_{gen}^{ratio}	P_{gen}^{avg} (kW)	P_{grid}^{avg} (kW)	η_{W2G} (%)
Reference	274	34,756	2.655	10.52	2.254	2.025	10.7
Grid Search P_{hydro}^{avg}	−20,000	48,000	3.088	11.58	2.444	2.235	11.8
Grid Search P_{grid}^{avg}	−3000	51,000	3.043	10.78	2.499	2.272	12.0
Gradient Descent	−5742	49,452	3.049	10.84	2.501	2.277	12.1

In Table 6, it is shown that, when using the Grid Search Method for maximizing P_{hydro}^{avg}, a maximum of 3.088 kW is absorbed by the point absorber. However, the average grid power in this case is 2.235 kW, which is lower than the Grid Search Method for maximizing P_{grid}^{avg}. Therefore, it can be concluded that maximizing the mechanical power captured by the point absorber does not necessarily lead to the maximum power at the grid side. The Gradient Descent Method identifies more detailed values for b^{RsdC} and k_c^{RsdC} in order to achieve P_{grid}^{avg} maximization, but the difference is negligible compared to Grid Search. The only significant advantage of the Gradient Descent in this case is the reduced simulation time of the algorithm to find these specific values.

3.3. Velocity Controller

The velocity controller (VelC) was first introduced in [28] for an array of DD WEC with energy storage at the DC link. The main idea behind the velocity controller is based on the equations given in (13).

$$\begin{cases} v_{pa}^{opt} = F_{exc}/[2R_i(\omega)] \\ P_{hydro} = F_{exc} \times v_{pa} \end{cases} \quad (13)$$

The first equation in (13) relates to the optimum velocity, v_{pa}^{opt}, the amplitude of the point absorber with the real part of the intrinsic impedance of the system, R_i, and the phase of the optimum velocity with the wave excitation force F_{exc}. Both the wave excitation force and the real part of the intrinsic impedance of the system are changing depending on the frequency of the incoming waves. The second equation in (13) shows that, in order to achieve maximum P_{hydro}, the F_{exc} and v_{pa} need to be in phase. The block diagram of the components comprising the velocity controller for DD WEC is shown in Figure 12.

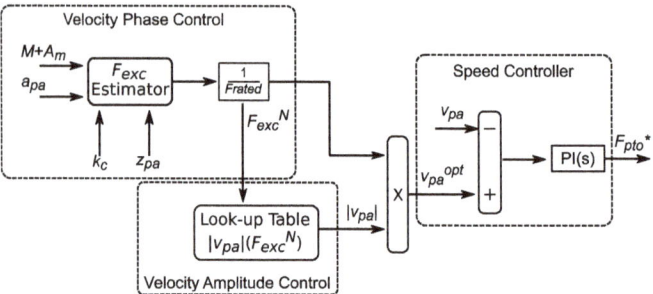

Figure 12. Block diagram of the velocity controller.

As is shown in Figure 12, the velocity controller is composed of three parts: Velocity Phase Control, Velocity Amplitude Control and the Speed Control. Velocity Phase Control uses the hydrodynamic parameters of mass, added mass and stiffness as well as the real time measurements of the point absorber acceleration (a_{pa}) and displacement in order to estimate F_{exc} (F_{exc}^{est}). The estimator uses Equation (14) to calculate F_{exc}^{est}.

$$F_{exc}^{est} = a_{pa} \times (M + A_m) + k_c \times z_{pa} \tag{14}$$

The F_{exc}^{est} is normalized by dividing it by the rated mechanical force F_{rated} to construct the phase component of F_{exc}. The damping component in the calculation of F_{exc}^{est} is not considered due to its minimal effect on the phase component. In addition, it is desirable to reduce the dependency of F_{exc}^{est} from the constant hydrodynamic parameters, which may change during the lifetime of the WEC. For this purpose, the use of Kalman filters may be considered for predicting the phase component of F_{exc}. The Velocity Amplitude Control of the velocity controller is mainly composed of a Look-up table, which generates the optimum velocity ($|v_{pa}|$) amplitude as a function of the phase component of $F_{exc}(F_{exc}^N)$. The amplitude and phase are multiplied to create the optimum velocity of the point absorber v_{pa}^{opt}. The v_{pa}^{opt} and the actually measured velocity v_{pa} are compared, and the velocity error is used as input to a PI controller. The output of the controller is the reference PTO force of the velocity controller, which aims to minimize the error between the optimum and the actually measured velocity.

Two sets of parameters are not predetermined in the velocity controller, the Look-up table and the tuning of the Speed Controller. The Look-up table can be defined in two different ways: the first assumes a linear relationship between the optimum velocity amplitude and the optimum phase. The linear relationship is based on the peak velocity the point absorber can have, and for simplification, the Look-up table can be replaced with a gain block with the peak velocity of the point absorber as a parameter. The second method assumes that the optimum velocity amplitude needs to be constant regardless of phase. The Look-up tables created based on these two different assumptions are shown in Figure 13.

The v_{linear} approach uses the maximum velocity of the point absorber (1 m/s) for all the normalized excitation force input points. Therefore, at rated excitation force, the optimum velocity will be the maximum, and a linear approach is used for the rest of excitation force input points. The $v_{constant}$ approach for the Look-up table aims at keeping the velocity of the point absorber near the maximum velocity, despite the low excitation force input. Therefore, at low F_{exc}^N, the velocity magnitude is high, and at high F_{exc}^N, the velocity magnitude tends to 1 m/s. The W2W WEC with the v_{linear} VelC is simulated under the same resource as the previous controllers, and the results are presented in Figure 14.

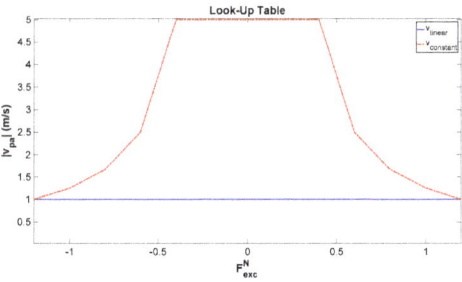

Figure 13. Two different approaches for the Look-up table of Velocity Amplitude Control.

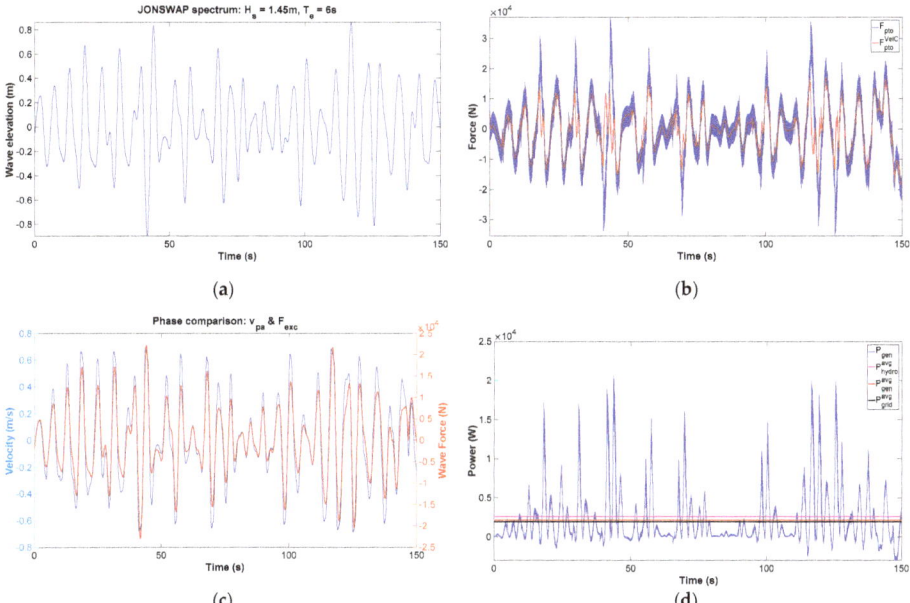

Figure 14. Simulation results using the v_{linear} VelC. (**a**) Wave resource used as input; (**b**) Reference and actual PTO force; (**c**) Phase comparison between the velocity and wave excitation force; (**d**) Instantaneous generator power and average hydrodynamic, generator and grid power.

In Figure 14, it can be observed that a good match is achieved between the reference and actual PTO force. This helps to produce a synchronized instantaneous velocity with the F_{exc}^N, which is an estimate of the input wave excitation force. In Figure 14c, this synchronization can be seen, and it can be observed that the instantaneous velocity of the point absorber follows the phase of the actual F_{exc} effectively. Figure 14d presents the power output at different stages of the W2W WEC. The peak instantaneous power of the PMLG is 20.5 kW, whereas the average values for the hydrodynamic power (P_{hydro}^{avg}), generator power (P_{gen}^{avg}) and grid power (P_{grid}^{avg}) are 2.5711 kW, 2.1479 kW and 1.9134 kW respectively. This leads to a η_{PMLG} = 83.5%, η_{Ele} = 89.1%, η_{hydro} = 13.6% and η_{W2G} = 10.2%. Finally, it is also shown that the VelC works in a similar way as a reactive controller which performs both phase and amplitude control. For that purpose, the PMLG absorbs power at some instances and in the 150 s simulation presented in Figure 14 the average negative power was 557 W. A summary of the results for the v_{linear} VelC and $v_{constant}$ VelC, along with all the options presented in this paper, is given in Section 3.4.

Power Maximization Process for the Velocity Controller

The main idea behind the power maximization process for the velocity controller is that the Look-up table can significantly affect the efficiency of the controller. In Section 3.3, by using a simple linear relationship between F_{exc}^N and $|v_{pa}|$, the phase between v_{pa} and F_{exc} was efficiently synchronized, but the amplitude of the velocity was not optimized. The gradient descent algorithm can be used to optimize the Look-up table of the VelC so that the W2W WEC system achieves maximum P_{grid}^{avg}. As was stated in the optimization process of the RdC and RsdC, the following aspects are taken into account for the optimization of the VelC:

- The objective is to increase the power exported to the grid, and not necessarily the power captured by the waves. The behavior of the electrical generator can change the losses in the electrical system.
- The mechanical and electrical constrains have to be included in the generation of the reference PTO force.
- Each WEC is installed in a specific wave climate. The Look-up table has to be modified to maximize power for the specific wave climate.
- Reactive power can bring the point absorber into phase with the wave excitation force for maximum power extraction from the waves, but may lead to excessive power consumption by the WEC. The amount of reactive power has to be controlled through optimization.

The optimized Look-up table for the 150 s wave resource shown in Figure 14a is given in Figure 15.

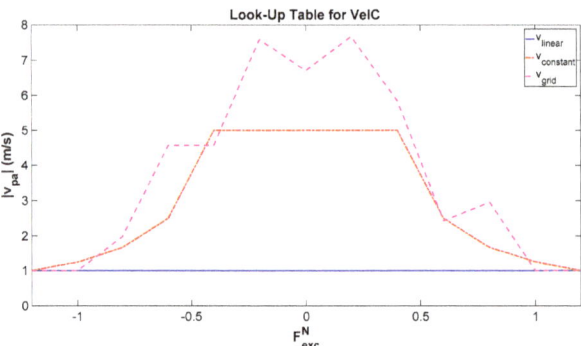

Figure 15. Optimized Look-up table, v_{grid}, of the VelC for maximum P_{grid}^{avg} in a specific wave climate. Comparison of v_{grid} with v_{linear} and $v_{constant}$.

The results from the simulation of the different velocity controllers are presented in Table 7. Based on the results presented, the VelC achieves high W2W efficiencies in all cases with smaller P_{gen}^{ratio} compared to RdC and RsdC keeping the control method cost efficient.

Table 7. Summarized Speed Controller results.

Controller	Variation	P_{hydro}^{avg} (kW)	P_{gen}^{avg} (kW)	P_{gen}^{ratio}	P_{grid}^{avg} (kW)	η_{W2W} (%)
VelC	v_{linear}	2.314	2.038	10.21	1.829	9.67
VelC	$v_{constant}$	3.214	2.623	9.56	2.460	12.99
VelC	v_{grid}	3.353	2.647	10.22	2.489	13.15

3.4. Summary of Controller Options

Table 8 summarizes the results obtained with the different controllers and compares the W2W efficiency between each controller and the RdC Reference. It is observed that the VelC v_{grid} has the highest W2W efficiency with an average of 2.498 kW for a 150 s simulation. The VelC v_{grid} achieved more than 5% extra efficiency, and an additional 1 kW power at the grid compared to the RdC Reference. High efficiencies were also achieved when the VelC $v_{constant}$, RsdC Gradient Descent and RsdC Grid Search were used.

Table 8. Summary of results for all the controller options presented.

Controller	Variation	P_{hydro}^{avg} (kW)	η_{hydro} (%)	P_{gen}^{avg} (kW)	P_{grid}^{avg} (kW)	η_{W2W} (%)	η_{W2W}^{Diff} (%)
RdC	Reference	1.915	10.10	1.691	1.470	7.77	0
RdC	Grid Search	2.007	10.53	1.747	1.565	8.26	+0.49
RdC	Gradient Descent	2.009	10.61	1.748	1.566	8.28	+0.51
RsdC	Reference	2.655	14.03	2.254	2.025	10.70	+2.93
RsdC	Grid Search	3.043	16.08	2.499	2.272	12.01	+4.24
RsdC	Gradient Descent	3.049	16.11	2.501	2.277	12.03	+4.26
VelC	v_{linear}	2.314	12.23	2.038	1.829	9.67	+1.99
VelC	$v_{constant}$	3.214	16.98	2.623	2.460	12.99	+5.22
VelC	v_{grid}	3.353	17.72	2.647	2.489	13.15	+5.38

4. Test Cases in a Realistic Environment

In the cases presented in Section 3, a complete simulation time of 150 s was used for the optimization of the controllers. However, WEC are expected to operate continuously for a much longer time. In this section, the controller variations with the highest W2W efficiency will be compared in a randomly generated sea state with the same significant wave height and energy period as the ones used for their optimization. In addition, low occurrence high-energy wave conditions will be examined in Section 4.2.

4.1. Dominant Operation Sea State

The realistic test case of the dominant operation sea state is based on the area of China around the Zhoushan Islands, which is presented in [32]. The average annual significant wave height is 1.45 m and the energy period is 6 s, which is representative of that area in China. The controllers tested in this section are optimized as described in Section 3 and are compared using a JONSWAP spectrum of the same significant wave height and energy period but randomized. Figure 16 presents the results of a 1500 s simulation for the RdC Gradient Descent, RsdC Gradient Descent and VelC v_{grid} controllers.

The results in Figure 16 show that the VelC requires significantly more negative power to increase the WEC efficiency compared to the RsdC. The average negative power for the VelC is 1.933 kW, whereas for RsdC it is 1.625 kW. The simulation results shown in Figure 16 are summarized in Table 9. It is noteworthy to mention that the P_{gen}^{ratio} of the VelC is smaller than for both the RdC and RsdC, making the VelC case a cost-effective alternative. In the simulations presented in Figure 16, the displacement of the buoy was always below 90% of the pa_{zmax}, and F_{ES} was zero in all cases. On the other hand, the power limitation system was enabled in the VelC v_{grid} only at 630 s, 733 s, 998 s and 1371 s, in order to keep the instantaneous power output below 30 kW. Despite the power limitation system being enabled, the VelC v_{grid} achieved the lowest P_{gen}^{ratio} and the highest W2G efficiency.

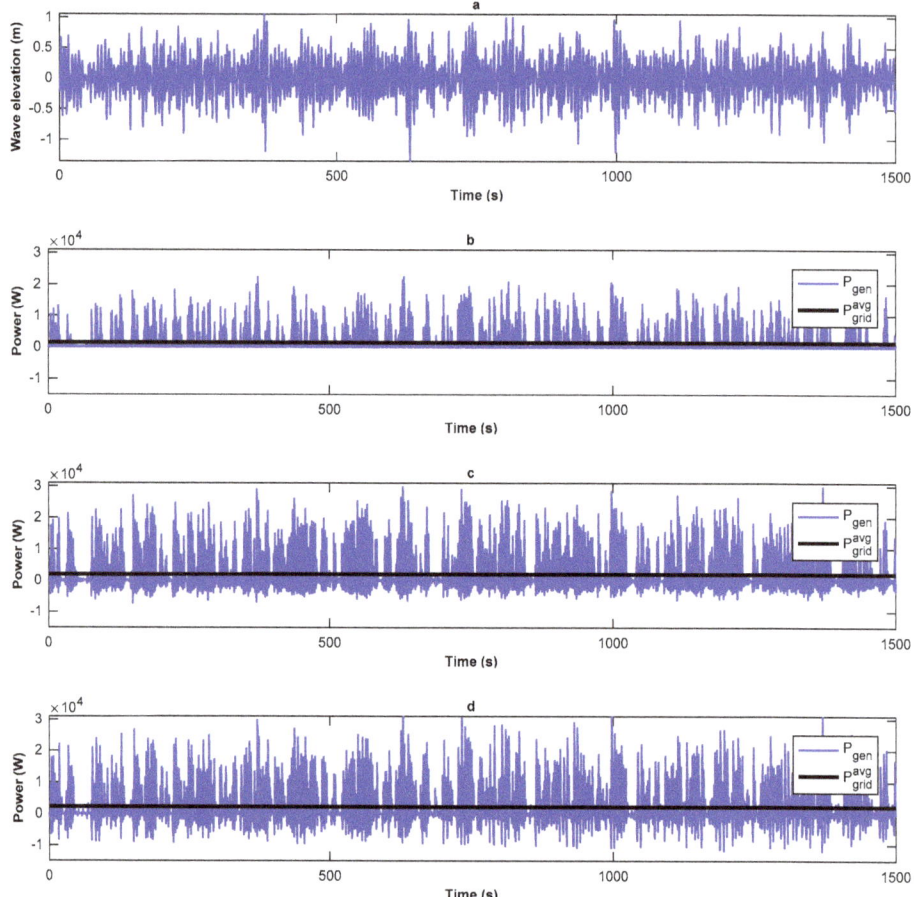

Figure 16. Results from a 1500 s simulation. (**a**) Wave resource input generated from JONSWAP spectrum with significant wave height 1.45 m and energy period 6 s; (**b**) Instantaneous generator power and averaged grid power for the RdC Gradient Descent case; (**c**) Instantaneous generator power and averaged grid power for the RsdC Gradient Descent case; (**d**) Instantaneous generator power and averaged grid power for the VelC v_{grid} case.

Table 9. Summarized results from the 1500 s random wave simulation for RdC, RsdC and VelC.

Controller	Variation	P_{hydro}^{avg} (kW)	η_{hydro} (%)	P_{gen}^{avg} (kW)	P_{gen}^{ratio}	P_{grid}^{avg} (kW)	η_{W2G} (%)	η_{W2G}^{Diff} (%)
RdC	Gradient Descent	1.954	10.32	1.708	13.05	1.514	8.01	0
RsdC	Gradient Descent	2.758	14.58	2.244	13.26	2.005	10.59	+2.58
VelC	v_{grid}	2.999	15.85	2.415	12.82	2.234	11.81	+3.80

Figure 17 compares the spectral density of the wave energy resource in which the controllers were optimized and the spectral density of the wave energy resource the controllers were tested in the realistic test case.

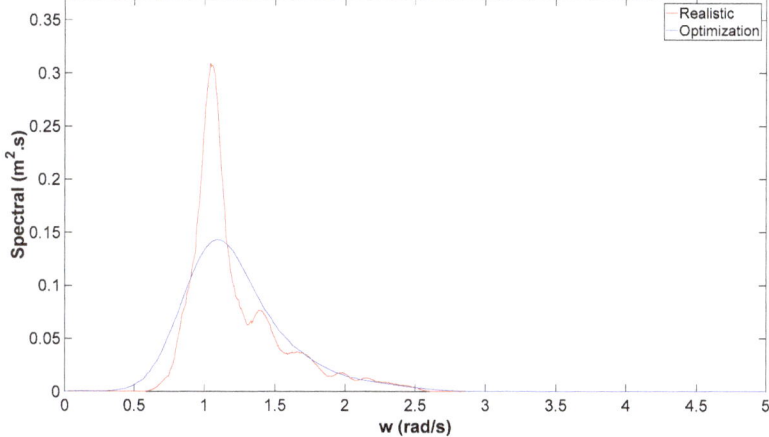

Figure 17. Spectral density of the wave resource used for the optimization of the controllers (blue line) and for the realistic simulation (red line).

It is observed that, despite the fact that the controllers were optimized in a wave energy resource that did not capture the peak spectral density, the efficiencies achieved in the 1500 s simulation were similar to those presented in Section 3, for which the controllers were tested in the resource they were optimized for. This shows that the controllers can efficiently operate in an environment that they were not optimized for, and that optimizing for one case does not restrict their ability to be efficient in a different wave energy resource.

4.2. High-Energy Sea State

In this Section the controllers are tested in a low occurrence high-energy sea state of the Zhoushan Islands. For the generation of the high-energy wave resource, the JONSWAP spectrum with significant wave height of 1.95 m and an energy period of 7 s is selected. This spectrum leads to a wave power potential per km 2.1 times larger than the dominant operation sea state. Figure 18 presents the 1500 s simulation results for the low occurrence high-energy sea state.

The results show a significant increase in the relative wave power generation in a high-energy sea state. The instantaneous wave power generation in all three cases, tested and shown in Figure 18, surpasses the 30 kW limit set by the power limitation mechanism discussed in Section 2.2. This demonstrates the difficulty of implementing a braking system only to limit the peak power generation in a WEC system that operates under high-energy wave conditions. Regarding the displacement constraint, the end-stop system was enabled a few times in all three cases, effectively limiting the displacement below pa$_{zmax}$. A summary of the results of the simulations presented in Figure 18 is given in Table 10. The peak-to-average ratio is kept a lot lower, almost half, compared to the averaged conditions tested in Section 4.1, and the efficiencies of all three controllers are significantly higher. Comparing the different controllers, it is observed that the trend is the same despite the fact that the controller settings are for averaged conditions. The VelC v_{grid} controller has the highest W2G efficiency and the lowest peak-to-average power ratio. The RsdC Gradient Descent Controller achieves a good efficiency, 0.78% more compared to the RdC Gradient Descent, but has the highest peak-to-average power ratio.

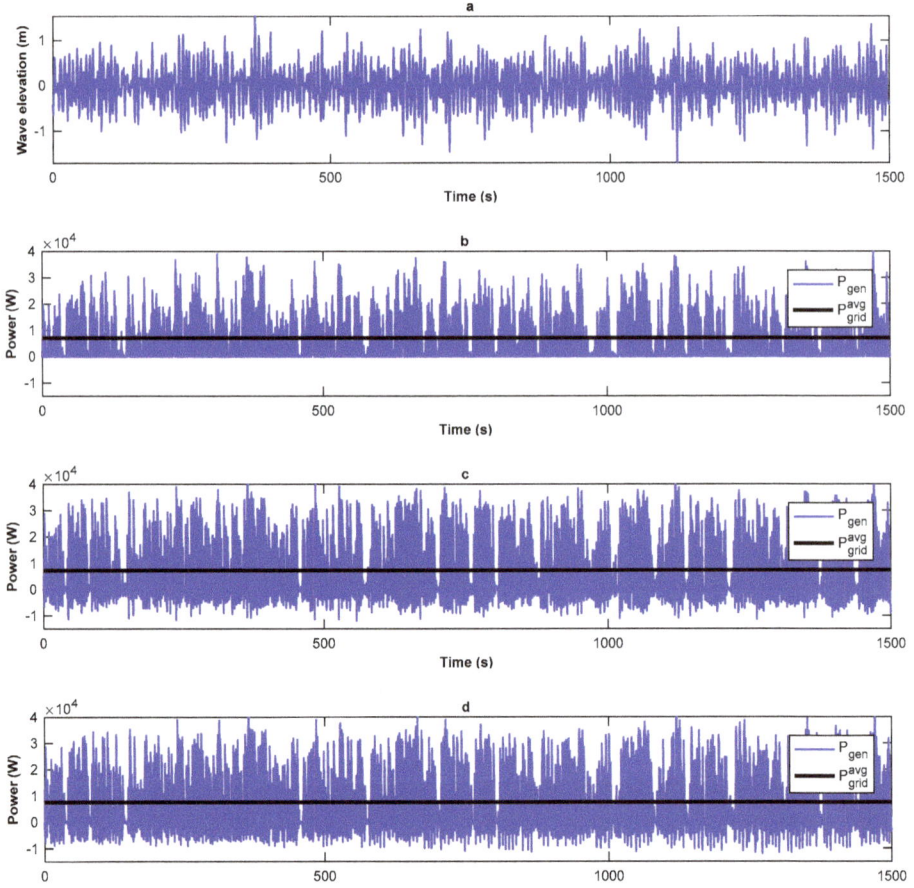

Figure 18. Results from a 1500 s simulation for a high-energy sea state. (**a**) Wave resource input generated from JONSWAP spectrum with significant wave height 1.95 m and energy period 7 s; (**b**) Instantaneous generator power and averaged grid power for the RdC Gradient Descent case; (**c**) Instantaneous generator power and averaged grid power for the RsdC Gradient Descent case; (**d**) Instantaneous generator power and averaged grid power for the VelC v_{grid} case.

Table 10. Summarized results from the 1500 s high-energy low occurrence wave simulation for RdC, RsdC and VelC.

Controller	Variation	P_{hydro}^{avg} (kW)	η_{hydro} (%)	P_{gen}^{avg} (kW)	P_{gen}^{ratio}	P_{grid}^{avg} (kW)	η_{W2G} (%)	η_{W2G}^{Diff} (%)
RdC	Gradient Descent	8.714	22.99	7.189	5.59	7.011	18.49	0
RsdC	Gradient Descent	9.969	26.30	7.533	5.61	7.305	19.27	+0.78
VelC	v_{grid}	10.114	26.68	7.739	5.48	7.559	19.94	+1.45

5. Conclusions

In this paper, a comprehensive method for optimizing three different W2G WEC controllers for maximum average power delivery to the connection point is presented. In addition to the optimization procedures and the WEC controller proposed in this paper, a full dynamic resource-to-grid hydrodynamic model is presented in detail, including power transmission and grid connection. A key contribution in this research paper is the proposal for a velocity controller as an alternative controller for WECs that does not require advance knowledge of the waves and frequency measurement of the incoming waves. The velocity controller requires the mass, added mass and linear restoring coefficient of the buoy in order to extract the optimum phase for the velocity. The amplitude of the velocity can be optimized for maximum power capture or maximum power to the grid. The velocity controller is compared with conventional controllers, the real damping controller and the reactive spring-damping controller. An optimization procedure for maximum power exported to the grid for the conventional controllers is also presented, and it shows that more power can be delivered with proper modification to the constant parameters of the reference PTO force generator. The simulation results show that the velocity controller can increase the efficiency of the W2W system more than 5% compared to the real damping controller with lower peak-to-average power ratio, increasing the cost-effectiveness of the components significantly. The RsdC also performs with increased W2G efficiency and the optimization process presented in this paper can improve its W2G efficiency by up to 2% with respect to the reference RsdC, and by more than 4% compared to the reference real damping controller. Finally, in the realistic simulation, it is shown that the optimized controllers can perform with high efficiencies in random environments, and without the need to be trained on a specific wave climate. Future research will focus on minimizing the electrical and mechanical stresses on highly energetic sea states by making the controller aware of the displacement and current constraints and a method to re-optimize the controllers presented during operation if the wave climate changes.

Author Contributions: Conceptualization, M.C.S.; Formal analysis, M.C.S.; Funding acquisition, J.S.; Investigation, M.C.S.; Methodology, M.C.S.; Project administration, J.S.; Supervision, J.S.; Visualization, M.C.S.; Writing—original draft, M.C.S.; Writing—review & editing, J.S.

Funding: The authors would like to acknowledge the support of the EPSRC in funding the work within this paper as part of the Joint UK–China Offshore Renewable Energy programme (EP/R007756/1).

Acknowledgments: The authors would like to acknowledge the support received for the hydrodynamic simulations by Richard Crozier and for the training of Edinburgh Wave Systems Toolbox.

Conflicts of Interest: The authors declare no conflict of interest. The funders had no role in the design of the study; in the collection, analyses, or interpretation of data; in the writing of the manuscript, or in the decision to publish the results.

References

1. Ruud, K.; Frank, N. *Wave Energy Technology Brief*; International Renewable Energy Agency (IRENA): Abu Dhabi, UAE, 2014.
2. World Energy Council. *World Energy Council World Energy Resources | 2016*; World Energy Council: London, UK, 2016; Volume 24.
3. Al Shami, E.; Zhang, R.; Wang, X. Point absorber wave energy harvesters: A review of recent developments. *Energies* **2019**, *12*, 47. [CrossRef]
4. Aderinto, T.; Li, H. Ocean Wave energy converters: Status and challenges. *Energies* **2018**, *11*, 1250. [CrossRef]
5. Mueller, M.A. Electrical generators for direct drive wave energy converters. *IEE Proc. Gener. Transm. Distrib.* **2002**, *149*, 446–456. [CrossRef]
6. Mendes, R.; Calado, M.; Mariano, S. Maximum Power Point Tracking for a Point Absorber Device with a Tubular Linear Switched Reluctance Generator. *Energies* **2018**, *11*, 2192. [CrossRef]
7. Penalba, M.; Ringwood, J.V. A high-fidelity wave-to-wire model for wave energy converters. *Renew. Energy* **2019**, *134*, 367–378. [CrossRef]

8. Balitsky, P.; Quartier, N.; Verao Fernandez, G.; Stratigaki, V.; Troch, P. Analyzing the Near-Field Effects and the Power Production of an Array of Heaving Cylindrical WECs and OSWECs Using a Coupled Hydrodynamic-PTO Model. *Energies* **2018**, *11*, 3489. [CrossRef]
9. Tedeschi, E.; Santos-Mugica, M. Modeling and Control of a Wave Energy Farm Including Energy Storage for Power Quality Enhancement: The Bimep Case Study. *IEEE Trans. Power Syst.* **2014**, *29*, 1489–1497. [CrossRef]
10. Tedeschi, E.; Molinas, M. Tunable Control Strategy for Wave Energy Converters With Limited Power Takeoff Rating. *IEEE Trans. Ind. Electron.* **2012**, *59*, 3838–3846. [CrossRef]
11. Sjolte, J.; Sandvik, C.M.; Tedeschi, E.; Molinas, M. Exploring the potential for increased production from the wave energy converter lifesaver by reactive control. *Energies* **2013**, *6*, 3706–3733. [CrossRef]
12. Ringwood, J.V.; Bacelli, G.; Fusco, F. Energy-Maximizing Control of Wave-Energy Converters: The Development of Control System Technology to Optimize Their Operation. *IEEE Control Syst.* **2014**, *34*, 30–55.
13. Li, B. Reaction Force Control Implementation of a Linear Generator in Irregular Waves for a Wave Power System. Ph.D. Thesis, The University of Edinburgh, Edinburgh, Scotland, 2012.
14. Shek, J.K.H.; Macpherson, D.E.; Mueller, M.A.; Xiang, J. Reaction force control of a linear electrical generator for direct drive wave energy conversion. *IET Renew. Power Gener.* **2007**, *1*, 17–24. [CrossRef]
15. Garcia-Rosa, P.B.; Ringwood, J.V.; Fosso, O.; Molinas, M. The impact of time-frequency estimation methods on the performance of wave energy converters under passive and reactive control. *IEEE Trans. Sustain. Energy* **2018**, *PP*, 1. [CrossRef]
16. Park, J.S.; Gu, B.-G.; Kim, J.R.; Cho, I.H.; Jeong, I.; Lee, J. Active Phase Control for Maximum Power Point Tracking of a Linear Wave Generator. *IEEE Trans. Power Electron.* **2017**, *32*, 7651–7662. [CrossRef]
17. Faedo, N.; Olaya, S.; Ringwood, J.V. Optimal control, MPC and MPC-like algorithms for wave energy systems: An overview. *IFAC J. Syst. Control* **2017**, *1*, 37–56. [CrossRef]
18. Fusco, F.; Ringwood, J.V. A Simple and Effective Real-Time Controller for Wave Energy Converters. *IEEE Trans. Sustain. Energy* **2013**, *4*, 21–30. [CrossRef]
19. Ammar, R.; Trabelsi, M.; Mimouni, M.F.; Ben Ahmed, H.; Benbouzid, M. Flux weakening control of PMSG based on direct wave energy converter systems. In Proceedings of the IEEE 2017 International Conference on Green Energy Conversion Systems (GECS), Hammamet, Tunisia, 23–25 March 2017; pp. 1–7.
20. O'Sullivan, A.C.M.; Lightbody, G. Co-design of a wave energy converter using constrained predictive control. *Renew. Energy* **2017**, *102*, 142–156. [CrossRef]
21. Falcão, A.F.O.; Henriques, J.C.C. Effect of non-ideal power take-off efficiency on performance of single- and two-body reactively controlled wave energy converters. *J. Ocean Eng. Mar. Energy* **2015**, *1*, 273–286. [CrossRef]
22. Genest, R.; Bonnefoy, F.; Clément, A.H.; Babarit, A. Effect of non-ideal power take-off on the energy absorption of a reactively controlled one degree of freedom wave energy converter. *Appl. Ocean Res.* **2014**, *48*, 236–243. [CrossRef]
23. Hansen, R.H. Design and Control of the PowerTake-Off System for a Wave Energy Converter with Multiple Absorbers. Ph.D. Thesis, Aalborg University, Esbjerg Aalborg, Denmark, 2013.
24. Wang, L.; Isberg, J.; Tedeschi, E. Review of control strategies for wave energy conversion systems and their validation: The wave-to-wire approach. *Renew. Sustain. Energy Rev.* **2018**, *81*, 366–379. [CrossRef]
25. Penalba, M.; Cortajarena, J.A.; Ringwood, J.V. Validating a wave-to-wire model for a wave energy converter-part II: The electrical system. *Energies* **2017**, *10*, 1002. [CrossRef]
26. Kovaltchouk, T.; Rongère, F.; Primot, M.; Aubry, J.; Ben Ahmed, H.; Multon, B. Model Predictive Control of a Direct Wave Energy Converter Constrained by the Electrical Chain using an Energetic Approach. In Proceedings of the European Wave and Tidal Energy Conference, Nantes, France, 6–11 September 2015.
27. Xiao, X.; Huang, X.; Kang, Q. A Hill-Climbing-Method-Based Maximum-Power-Point-Tracking Strategy for Direct-Drive Wave Energy Converters. *IEEE Trans. Ind. Electron.* **2016**, *63*, 257–267. [CrossRef]
28. Sousounis, M.C.; Gan, L.K.; Kiprakis, A.E.; Shek, J.K.H. Direct drive wave energy array with offshore energy storage supplying off-grid residential load. *IET Renew. Power Gener.* **2017**, *11*, 1081–1088. [CrossRef]
29. Crozier, R. RenewNet Foundry 1.6. Available online: https://sourceforge.net/projects/rnfoundry/ (accessed on 18 March 2019).

30. Polinder, H.; Slootweg, J.G.; Hoeijmakers, M.J.; Compter, J.C. Modelling of a linear PM machine including magnetic saturation and end effects: Maximum force to current ratio. *IEEE Trans. Ind. Appl.* **2003**, *39*, 1681–1688. [CrossRef]
31. Wu, B.; Lang, Y.; Zargari, N.; Kouro, S. Power Converters in Wind Energy Conversion Systems. In *Power Conversion and Control of Wind Energy Systems*; John Wiley & Sons, Inc.: Hoboken, NJ, USA, 2011; pp. 87–152.
32. Wan, Y.; Fan, C.; Zhang, J.; Meng, J.; Dai, Y.; Li, L.; Sun, W.; Zhou, P.; Wang, J.; Zhang, X. Wave energy resource assessment off the coast of China around the Zhoushan Islands. *Energies* **2017**, *10*, 1320. [CrossRef]

© 2019 by the authors. Licensee MDPI, Basel, Switzerland. This article is an open access article distributed under the terms and conditions of the Creative Commons Attribution (CC BY) license (http://creativecommons.org/licenses/by/4.0/).

Article

Wave Power Output Smoothing through the Use of a High-Speed Kinetic Buffer

Brenda Rojas-Delgado [1], Monica Alonso [1,*], Hortensia Amaris [1] and Juan de Santiago [2]

[1] Electrical Engineering Department, Universidad Carlos III de Madrid, 28911 Madrid, Spain; brenda.rojas@alumnos.uc3m.es (B.R.-D.); hortensia.amaris@uc3m.es (H.A.)
[2] Division of Electricity, Department of Engineering Sciences, The Angstrom Laboratory, P.O. Box 534, SE-75121 Uppsala, Sweden; Juan.Santiago@angstrom.uu.se
* Correspondence: monica.alonso@uc3m.es; Tel.: +34-91-624-8433

Received: 13 April 2019; Accepted: 6 June 2019; Published: 10 June 2019

Abstract: In this paper, a new control strategy for power output smoothing in a hybrid wave energy installation coupled to a flywheel energy storage system (FESS) is proposed. The control scheme is composed by three stages: a wave generator clustering process at the farm connection point; a power filtering process; and the control of the flywheel energy storage in order to improve the power output of the hybrid wave farm. The proposed control is validated at the existing Lysekil Wave Energy Site located in Sweden, by using real generator measurements. Results show that the application of the flywheel energy storage system reduces the maximum peak power output from the wave energy installation by 85% and the peak/average power ratio by 76%. It is shown that the proposed system can reduce grid losses by 51%, consequently improving the energy efficiency of the power network. The application of the proposed control strategy allows the hybrid wave power plant to follow a power reference signal that is imposed by the grid operator. In addition, the study demonstrates that the application of the proposed control allows the hybrid wave power plant to follow a power reference signal that is imposed by the grid operator. In addition, the study demonstrates that the application of the proposed control enables a wave farm with flywheel energy storage to be a controllable, flexible resource in order to fulfill future grid code requirements for marine energy installations.

Keywords: wave energy; energy storage; flywheel; power take off (PTO); flywheel energy storage system (FESS)

---

1. Introduction

In 2016, the World Energy Council stated that the global marine generation capacity was 92 PWh/year, which points to the possibility that wave energy could become more competitive compared with other forms of renewable generation in the future [1]. Nevertheless, despite its viability, the costs are still high. The installation and operation of grid-connected wave energy plants still need to overcome technological barriers and gaps in important knowledge to meet economy of scale criteria. Apart from the problems derived from generation unit costs, wave energy sources also face a variety of environmental, infrastructural, and socioeconomic obstacles, such as the uncertainty of grid-connected infrastructures, obstacles related to integration with the electricity market, and power quality problems.

One of the most important barriers to effective wave energy integration with power networks is the high variability in the power generated by wave energy sources. It must be highlighted that these energy installations are located in coastal areas, which generally have weak grids that necessitate strategies to manage their associated issues; such actions include voltage regulation, frequency leveling, power factor correction, and harmonic mitigation. In addition, these wave energy installations lack specific grid code technical requirements that define their specific power capabilities. Consequently, the existing grid codes have proved to be inadequate for wave energy installations [2].

The aforementioned barriers show that there is a need for efficient control tools to enhance the regulation of power output by wave energy sources [2]. Comparatively few research has been dedicated to power smoothing at wave energy installations. Most of the research publications related to marine installations have focused on the design of the wave front-end interface [3–5], the study of the transmission technology [6], or the mechanical design [7–11].

Very few techniques have been proposed to improve the wave power output, and these can be classified on the basis of whether they use an electrical energy storage system (ESS) to accomplish their goal [12]. Some studies have proposed clustering a set of wave energy converters (WECs) in order to generate power that can be injected into the grid with reduced oscillations without needing to use an ESS. Power output aggregation also improves the on-shore connection system [13]. Other studies have used ESSs to mitigate short-term voltage fluctuations. However, in these applications, it has been found that the rating of the ESSs increases as the installed WEC power increases [14–16]. In [17], supercapacitors were used to improve wave farm power output at laboratory installations. However, only regular waves were considered, and the behavior of synthetic regular waves differs greatly from real wave conditions. From this aspect, it must be noted that the use of ESSs has proved to be effective for smoothing wave power output. In spite of this, the use of flywheels in marine energy installations is very rare, mainly because it is not easy to coordinate the operation of wave energy generators by controlling the energy stored in the flywheel.

The contributions of this paper are threefold:

- The development of control methods for power output regulation at hybrid wave energy installations combined with a flywheel is presented. The proposed control strategy can be used for wave power output regulation in order to satisfy future grid codes for marine energy sources.
- The proposed control considers the application of flywheel ESSs for wave energy plants to smooth the power output delivered to the grid.
- It is worth highlighting that the developed control strategies were applied using data from an existing wave emplacement, namely, the Lysekil Wave Energy Site located in Sweden.

The proposed control is able to improve not only the power injected into the grid but also the power following capabilities of the wave power installation. The modeling and control were validated at the experimental wave power plant located at the Lysekil Wave Energy Site developed by Uppsala University in Sweden. To this end, data gathered from the real emplacement were later utilized to implement the developed control algorithms.

2. Wave Energy Characteristics

Wave energy resources are three times the size of currently available resources and almost double the energy density, and they offer higher forecast accuracy when compared with wind and solar renewable sources [3]. With respect to the annual production ratio, WECs work 90% of the hours of the year; this ratio is significantly greater than the 20–30% of other renewable generation technologies, such as solar and wind [12]. It is estimated that 10% of the energy consumed by the European continent could be supplied solely by marine energy by the year 2050. However, despite the advantages that wave energy presents and the availability of its resources, it is necessary to overcome a series of challenges: to become a competitive technology, a great investment in marine energy development must be made so that the installation costs can be reduced, grid integration can be optimized, and grid code requirements can be met [18].

- Wave energy is characterized by the height and frequency of the waves, which experience constant variation; as a result, a marine installation's power output does not comply with the quality criteria required for electric grids. To improve the power quality of the delivered energy, it is possible

to install storage systems in marine emplacements or even group wave generators in a way that allows oscillations of the power output of some units to be compensated by the power output of other units in the same cluster.
- Wave direction is very changeable, and for this reason, it is necessary to carry out exhaustive studies to determine the optimal location of the marine generation units' mooring systems.
- With respect to grid connection requirements, there are no grid codes regulating the coupling of marine generation emplacements; therefore, the codes that have been used for wind installations are applied instead. However, the variability in the power output that marine installations inject into the grid must be considered because their behavior differs greatly from other installations.

Figure 1 shows an example of the power output that corresponds to the permanent magnet linear generator WEC installation at the Lysekil Marine Platform in Sweden. In this case, there are three WEC units that are formed by a surface buoy connected to a linear generator moored to a foundation on the seabed [19–21]. It can be seen in Figure 1 that in a one-minute time interval, the power output of the marine installation shows great power oscillations. As it can be noted, marine production presents a peak power about 17 kW, which is equivalent to more than 5.8 times the mean power delivered to the grid, i.e., 2.88 kW. There is great irregularity found in a WEC's power production as a consequence of the variability in wave natural energy resources, and this irregularity makes the grid integration of this technology difficult to achieve.

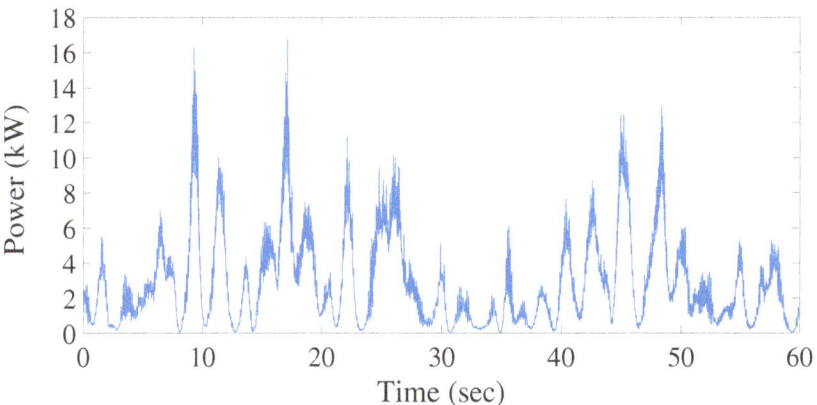

Figure 1. WEC's power output for the Lysekil wave farm.

3. Wave Power Output Signal Improvement

Converting wave energy into electricity involves three fundamental steps: wave resource absorption, wave power transmission, and electricity generation (Figure 2). However, wave power is a non-controllable and intermittent energy source. This huge variability in the power output attributed to wave energy emplacements necessitates the implementation of a fourth step with respect to the conversion process. The application of signal conditioning devices in this state is particularly crucial in installations that use direct-driven wave energy converters. The reason for this is that there is no mechanical stage between wave resource absorption and electricity generation that can absorb the oscillations caused by wave motion.

To improve the wave power output, several techniques have been implemented, and these can be classified on the basis of whether they use electrical energy storage systems to accomplish their goal [12].

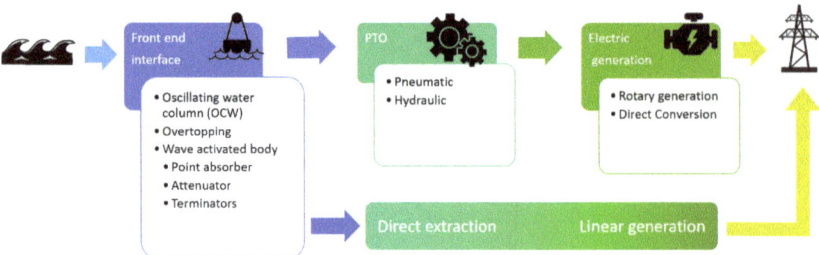

Figure 2. Transformation of wave resources into electrical energy.

Techniques that do not employ storage systems to improve the power output include the following:

- The first approximation consists of clustering a set of WECs at the same connection point (PCC) in order to generate power that can be injected into the grid with reduced oscillations. This power output aggregation also improves the on-shore connection system [22].
- Another option is the control of the wave power converter. The active control of the power take-off (PTO) system can improve the power output of a grid-connected marine emplacement [23].
- In those cases in which hydraulic PTOs are used, it is possible to reduce the power output oscillations by means of water accumulators at sea level [24]. The system utilized by overtopping reservoirs allows for a small power output regulation. Nevertheless, storing wave energy in its hydraulic form in other types of installations could lead to extra installation costs [25].

Among ESSs that have been used to mitigate short-term voltage fluctuations are the following examples:

- Capacitors were used in [26], in which a power converter controlled the charge/discharge operation to smooth the power output oscillations of the marine installation. It was demonstrated that the size of the capacitor depends on the size of the installed WEC power. Consequently, its use is restricted to small wave energy plants.
- In [17], supercapacitors were used to improve the wave farm power output of laboratory installations, and synthetic regular waves were used as input signals (see Figure 3).

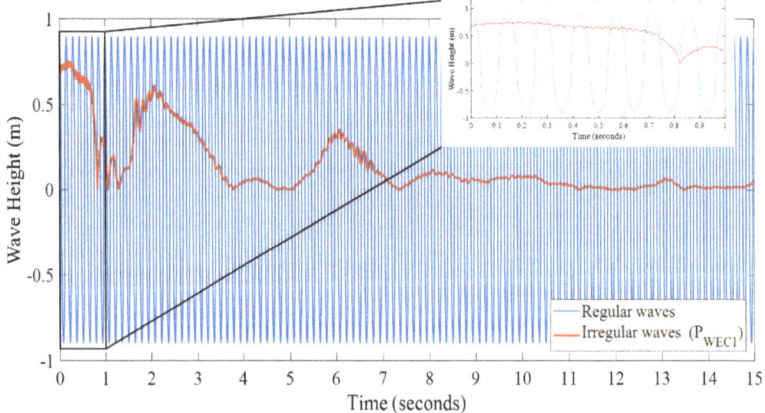

Figure 3. Regular and real (irregular) waves' waveform comparison.

Energy storage systems coupled with wave installations must handle very a fast reaction time and must be able to cope with hundreds of charge/discharge cycles per day [27].

4. Flywheel Energy Storage Systems

Power systems are currently experiencing a high penetration of small stochastic distributed generation units. Energy storage systems present themselves as promising alternatives for addressing the great variability associated with the power output variability and power quality delivered to the grid. This is because ESSs can either deliver a great amount of energy in short time periods or provide energy to the power systems over longer periods [28]. According to [29], there are at least 11 types of energy storage systems with different specifications and different market shares. Overall, batteries are the cheapest and most extended storage system, but flywheels have emerged as a very competitive technology. Power electronic converters have been applied extensively in wind power installations [30]. They can be used for controlling variable-speed wind turbines and also for improving the reactive power capability at the connection point [31]. The use of power electronic converters for power quality applications is normally related to voltage waveform quality [32] and the continuity of supply, but they are not able to perform power smoothing without an additional storage system. This is an important distinction from a flywheel energy storage system (FESS), which can be used for either power quality purposes (as can be performed by power electronic equipment) or power output smoothing (which cannot be performed by power electronic equipment). An FESS consists of a rotational mass that is capable of storing an amount of kinetic energy that is directly proportional to the mass and the square of its rotational speed. For energy production purposes, these rotational masses either store or deliver electricity by means of converting the kinetic energy into electrical energy when coupled with a motor or generator.

The main components of an FESS are as follows [33] (Figure 4):

- The primary component is a rotating mass that is commonly known as a flywheel. The rotating mass is able to store kinetic energy, so the bigger the size of the flywheel, the greater the stored energy. The stored kinetic energy is strongly dependent on the mass rotational speed.
- An electrical machine is in charge of transforming the mechanical energy to electrical energy or vice versa (generator/motor).
- Power electronics are used for controlling the performance of the electrical machine.
- Magnetic bearings hold the flywheel's weight.
- An external inductor is used for improving the Total Harmonic Distortion (THD) that is created by the generators of the permanent magnets in the electrical machine.

Several types of electrical machines can accompany the flywheel when the charge/discharge process takes place. These types of machines include the permanent magnet synchronous machine (PMSM) [34], brushless DC electrical machine (BLDCM) [35], induction machine (IM) [36], switched reluctance machine (SRM) [37], homopolar machine (HM) [38], synchronous machine (SM) [39], or bearingless electrical Machine (BM) [40]. Rotor bearings are classified as permanent (Passive) magnetic (PMB), active magnetic (AMB), and superconducting magnetic (SMB). Mechanical bearings have practically fallen into disuse since the implementation of their magnetic counterparts.

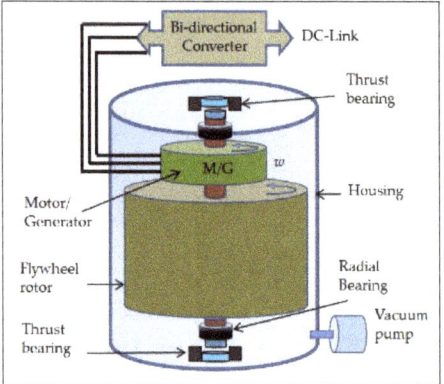

Figure 4. Structure and components of a flywheel. (Reproduced with permission from the authors [33]).

In general, FESSs are connected to the grid by means of power converters that help to control the energy storage operation. The back-to-back topology represents a scheme in which there is an AC/DC converter after the grid that is coupled with a flywheel storage system, and there is also one placed after an AC/DC converter. This topology is seen in [41].

Flywheels also have several applications aside from merely ensuring power quality and load management, as seen in Figure 5 [27,42–44]. Flywheels have high standby losses. These are due to friction losses and are unavoidable. This is the reason that flywheels are only applied in cases of frequent charge/discharge cycles. However, this disadvantage is only relevant for applications with long standby periods.

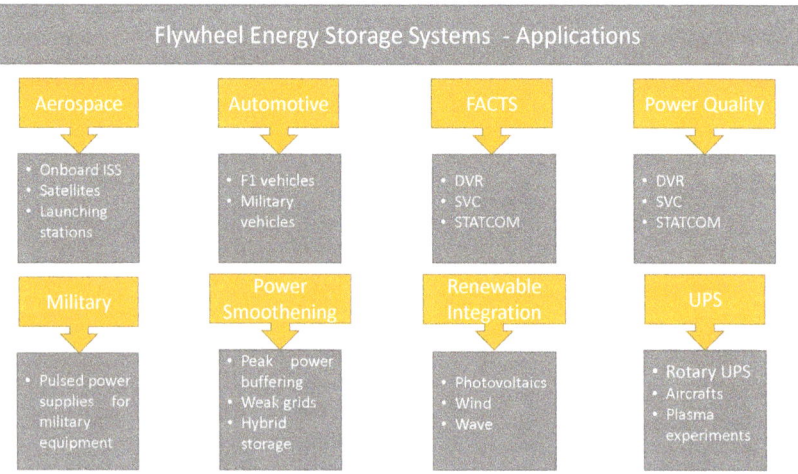

Figure 5. Applications of Flywheel Energy Storage Systems [27,42–44].

Flywheels present some advantages when compared with electrical batteries:

- First, flywheels do not require chemical components, so they have less impact on the environment and are considerably low-maintenance [27,29].
- Moreover, flywheels have a higher power and energy density [27].
- They also have a shorter charging time. This translates to a power output that can come from primary sources with high variability in a short period of time, such as wind and wave power [27].

- Furthermore, they have higher efficiency (90–95%) [33]: for the same number of cycles, advanced Lead–Acid batteries have an efficiency between 75 and 90%, whereas Li-ion batteries register an efficiency of around 87% [29].
- Flywheel ESSs are cheaper compared with other batteries storage systems, such as Li-ion and flow batteries [45], and because their performance does not degrade over time, they do not need to be replaced, contrary to the requirements of battery devices. In general, the economics of flywheel units is superior compared with battery technologies; for example, for one cycle per day, the battery cost of a Li-ion and flow battery can be up to 14.2% and 24% higher than the flywheel cost, respectively. This difference increases for three or more cycles per day, as shown in [45].

From the construction point of view, flywheel systems installed with WECs as part of an underwater installation require additional material treatment because they can be affected by water corrosion [46]. Moreover, two of the most important challenges in flywheel development are the weight and safety rotational speed, which limits its maximum energy capacity. Although implementation costs increase as the flywheel size increases, the initial construction costs can be offset by relevant savings associated with grid ancillary services, such as load shifting, demand response, congestion management, or energy savings.

4.1. Flywheel Components

Figure 4 depicts a typical flywheel's structure and components. Three substructures are clearly differentiated, namely, a vacuum pump, a DC link, and flywheel housing. The vacuum pump regulates the removal of air inside the flywheel housing, and the DC link is connected to the motor/generator inside the housing by means of a bidirectional AC/DC converter.

The flywheel housing has several components, namely, the flywheel rotor and the motor/generator, which converts the kinetic energy stored in/delivered by the flywheel rotor into electrical energy. This energy is later sent to the DC link when it is working in generator mode. Both rotating masses are connected by means of their central shaft, whereas the electrical connection between the housing and rotating components is achieved by both thrust and radial bearings, which can be brushless or non-brushless depending on the technology.

4.1.1. Flywheel Rotating-Mass

The flywheel's rotating mass is the key component of the FESS, and it consists of a disc capable of storing electrical energy in the form of kinetic energy. This energy storage makes it possible to smooth the power output of the installations, which are characterized by the high variability in a natural energy resource. The selection of the disc material is extremely important for establishing the operative characteristics of the device. Steel was used to manufacture flywheels when they were first fabricated, but this material presents stress limitations. In the 1970s, composite materials started being utilized for flywheel manufacturing because they can work at high speeds and support a large degree of the fatigue induced by rotational stress. For this reason, small, composite-based flywheels are used for functionalities that require performance speeds that are above 10,000 rpm, while those made of steel are utilized for applications that do not require such high speeds (up to 10,000 rpm).

Power and energy are decoupled in flywheel storage systems. The power rate is defined by the electrical machine that is coupled to the flywheel. Kinetic energy, E_k, is defined by the moment of inertia J and the square of the rotational speed ω, as shown in (1).

$$E_k = \frac{1}{2}J\omega^2 \quad (1)$$

From (1), it can be deduced that to increase the kinetic energy stored in the flywheel, actions need to be applied to change its moment of inertia or its angular speed. The moment of inertia is dependent on the mass and the disc ratio of the flywheel, as given in (2). Moreover, because the cylinder volume is directly proportional to the ratio r and the height h, the moment of inertia can be calculated according to (3), where ρ_m is the material density of the flywheel's rotating mass.

$$J = \frac{1}{2}mr^2 \qquad (2)$$

$$J = \frac{1}{2}\rho_m \pi h r^4 \qquad (3)$$

Similarly, useful energy can be obtained in two different ways. The first way is by using the difference between the maximum and minimum rotational speeds, as defined in (4). The second way, which is given in (5), is based on the moment of inertia's decomposition for a hollow cylinder, where m is the cylinder mass, r_{outer} is the outer radius, and r_{inner} is the inner radius [33].

$$E_k = \frac{1}{2}J(\omega_{max}^2 - \omega_{min}^2) \qquad (4)$$

$$E_k = \frac{1}{2}m\omega^2(r_{outer}^2 - r_{inner}^2) \qquad (5)$$

4.1.2. Magnetics Bearings

Magnetic bearings are located inside the FESS cavity, especially at the upper and lower parts of the disc that contains the flywheel's rotating mass. The main function is to hold the weight of the flywheel disc, so it can thus be considered the main mechanical component of the FESS. The flywheel disc is maintained between the two magnetic bearings by the repelling electromagnetic forces that are generated inside the housing, and there is no need for them to be in contact with the disc. For this reason, flywheels that employ magnetic bearings have less frictional force than their mechanical counterparts; in addition, they can reach a higher rotational speed and have a longer life cycle and capacity with minimal maintenance [28]. To limit the friction of the elements contained in the housing, the housing is filled with low-pressure air or helium [47]. The main inconvenience with magnetic bearings is their control mechanism. When there are failures in the control system or overloads, it is necessary to have an additional mechanical bearing system.

There are currently three types of magnetic bearings: passive, active, and superconducting.

- Passives: The main advantages of these elements are their low friction losses and low cost [48]. However, they are used as backup bearings because their performance presents instabilities.
- Actives: These bearings are used as additional units to reduce rotor vibrations. Their main advantage is that they present higher controllability and longer life cycles than passive ones, but active magnetic bearings do present higher losses due to the existence of bias currents. Combining active and mechanical bearings is a good choice in terms of controllability, stability, viability, and costs [41].
- Lastly, superconducting magnetic bearings are the most widely utilized type in applications that require high speed since they have more stability, longer life cycles, and lower losses. The main inconvenience is their very low operating temperature, which requires the addition of a cryogenic system that can provide these low temperatures. Consequently, the cost of the equipment increases [41].

4.1.3. Electrical Machine

The electrical machine plays a key role in FESSs because it acts as an electromechanical device in charge of establishing the energy interchange between the flywheel and the grid. At the same time, it converts electrical energy into mechanical energy and vice versa. In other words, the electrical machine works as a generator when the electric grid demands energy from the flywheel, in which case the speed of the flywheel is reduced; when the machine needs to act as a motor to store energy in the flywheel, its acceleration is increased. Permanent magnet synchronous machines (PMSGs) are the most commonly utilized electrical machines in FESSs because they present lower rotor losses, higher energy densities, and higher efficiency compared with other electrical machines. They are used in applications that require rotational speeds of about 50,000 rpm, but there are current developments that can reach ultrahigh velocities (150,000–300,000 rpm) [49].

4.1.4. Power Converters

Figure 6a shows the basic electrical scheme of a flywheel device. It can be observed that the flywheel disc is connected to an electrical machine that performs the electromechanical conversion. The electrical machine is connected to the grid by a back-to-back converter. In general, the flywheel's electronic converters perform an AC/DC/AC conversion, with the back-to-back configuration being the most commonly used for FESSs [41]. The grid-side converter converts the AC voltage to DC. After the DC stage, the machine-side converter converts the DC voltage back to AC. Because the electrical machine works as a generator or a motor, grid-connected converters are bidirectional. The machine-side converter is in charge of controlling the rotational speed and the power interchange by means of active power regulation. On the other hand, the grid-side converter takes care of maintaining the DC-link voltage at constant levels. When the electrical machine works as a motor, the machine-side converter works as a power inverter, while the grid-side converter works as a rectifier. When working as a generator, the machine-side and grid-side converters work as a rectifier and inverter, respectively. When the flywheel is utilized to improve the power output of the DG units, which can be wind or wave energy, the flywheel can use the machine-side converter only, and it is connected to the DC link of the back-to-back DG converters (see Figure 6b). In this case, the grid-side converter should account for the rated capacity of both the DG and the flywheel [41].

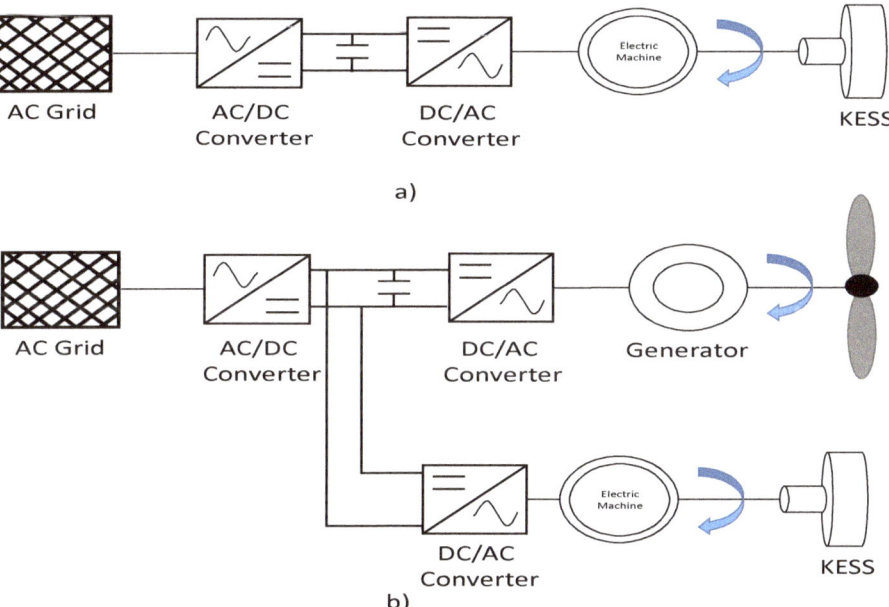

Figure 6. Electric schemes of a grid-connected flywheel. (**a**) back-to-back topology (**b**) back-to-back wind-FESS combined topology [41].

5. Lysekil Research Site

5.1. Location

According to the classification established by [50], the Lysekil wave farm (Lysekil Research Site, LRS) can be included in the group of the last "pre-commercial stage gate requirements test sites". The installation is located on the west coast of Sweden, 100 km from the southwestern part of Göteborg city and 2 km from the coast of Lysekil city. The location encompasses a surface of 40,000 m^2 with coordinates of 58° 11'850" N 11° 22'460" E and 58° 11'630" N 11° 22'460" E. The sea bottom is located at a depth of 24–25 m, and optimal conditions are created for mooring WEC devices [51]. The location of the LRS is surrounded by islands; this allows the wave farm to have a good sea state, even in winter, with an energy density of 2.6 ± 0.3 kW/m [52]. Figure 7 shows the location of the installation and the sea state in the surroundings of the LRS. Figure 8 shows the wave climate matrix that corresponds to the Lysekil wave farm location.

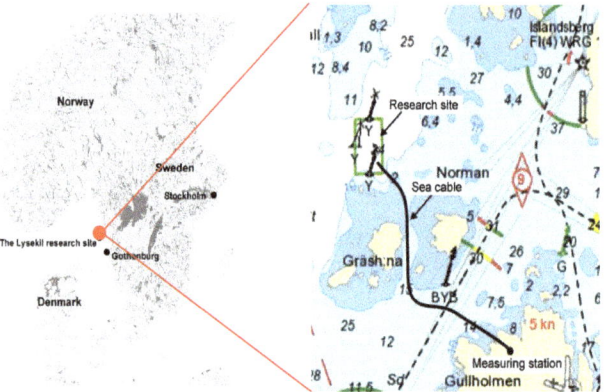

Figure 7. Location and state of the sea at the Lysekil marine facility [51,53].

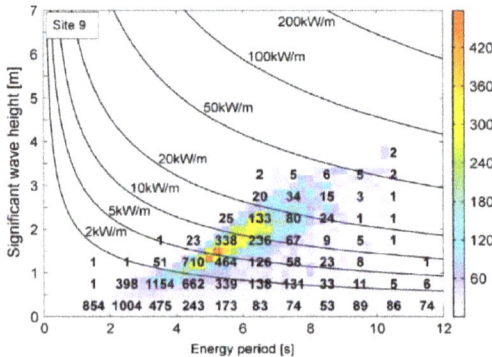

Figure 8. Wave climate matrix at Lysekil research site [54].

The LRS started its operation in 2003, and the first measurements related to its sea states were taken in 2004. In March 2006, the LRS installed the first buoy, known as L1; a 2.9-km-long wire connects it to the measurement substation that is located on the Hermanö island at the southern point of the LRS. In 2009, two additional WEC devices, namely, L2 and L3, were installed [55]. Together with the WECs, 21 environmental buoys were installed in 2007; the intention was to measure the impact of the WEC units on the marine surroundings of the installation and vice versa. Between 2009 and 2010, five new

WEC units (L4, L5, L7, L8, and L9) were installed, and three more WEC units (L10, L11, and L12) were installed in 2015. As of November 2015, all of them were grid-connected, with an installed capacity of 200 kW [55]. To date, more than ten WEC models have been analyzed at the LRS [56]. In the summer of 2008, the LRS was completed with the installation of a measurement tower, which is situated 150 m away from the LRS. The measurement tower includes a camera that makes it possible to correlate the wave motion captured in the images with the installation's power output. Figure 9 lists (from left to right) the components in the LRS scheme and the scheme itself: namely, the observation tower, the biological buoy, the buoy for the measurement of the sea state by means of wave characteristics, the WEC coupled with its energy absorber buoy and the linear generator for electricity conversion, the marine substation, and the off-shore measurement station.

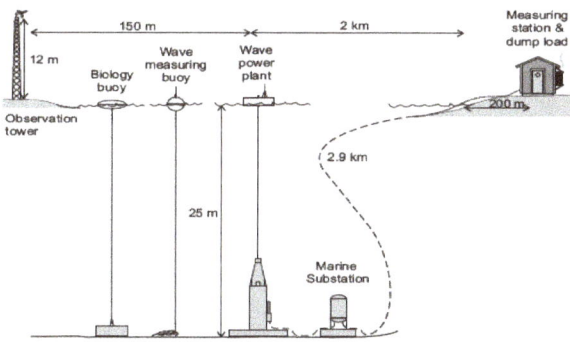

Figure 9. Outline of the basic components of the LRS [57].

5.2. Technology Used

In the LRS, point absorbers are used to extract energy from the waves, while the conversion of wave energy into electrical energy is performed using a direct-driven linear generator (DDLG). The movement of the waves is transmitted from the point absorber to the generator translator through a cable. To improve the performance of the generators in extreme wave conditions, WEC units are provided with springs in their upper part that limit the translator motion inside the DDLG's cavity. Because the LRS is a research installation, several DDLGs (Figure 10a) and point absorber buoys (Figure 10b) have been tested [19].

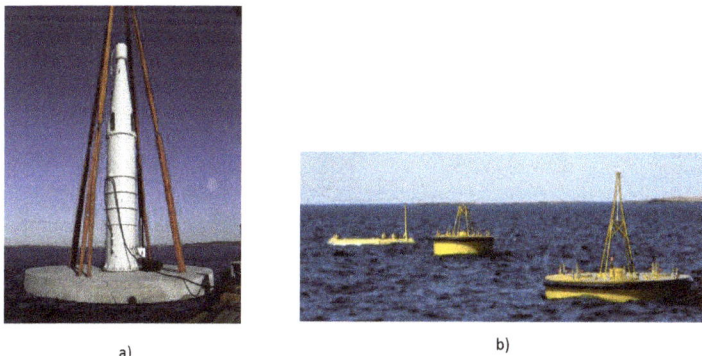

Figure 10. DDLG (**a**) and point absorber (**b**) buoys at the LRS [19,51].

The technical characteristics of both the L1 and L9 models are described in [51].

5.3. Marine Substation

The LRS installation is grid-connected through a marine substation, with a rated power of 96 kVA, and a 2.9-km-long wire (see Figure 11) that is used for the configuration of the first WEC units [53].

The wave power generators are provided with a 12-Ω dump load, which handles the dissipation of the energy delivered by the WEC units when they are disconnected from the substation.

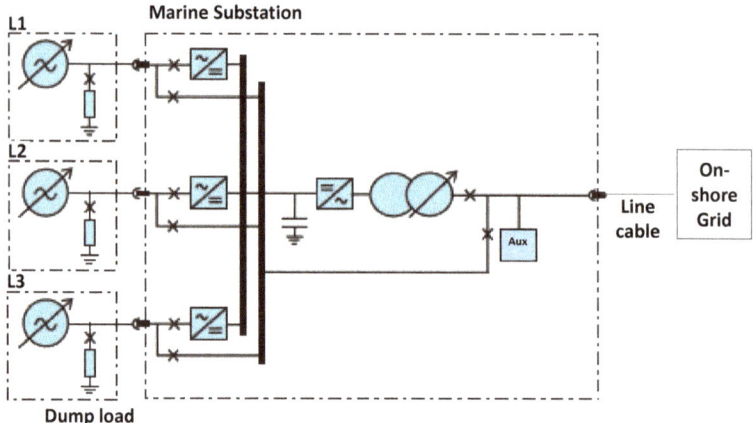

Figure 11. Electrical diagram of the LRS on-shore connection [58].

The power output of each WEC unit is rectified by a six-diode bridge converter. In the marine substation's DC stage, the WEC units are interconnected in parallel. The electrolytic aluminum capacitors in the DC stage allow for the smoothing of the output signal that is sent from the installation to the inverter. The AC voltage at the output of the inverter is increased to the grid voltage (1000 V). This is done using a Y-Y power transformer that is provided with tap regulation (80-100-125-250/1000 V) and a tap-changing mechanism composed of off-load circuit breakers. The elements that make up the substation are confined within a 3 m^3 nitrogen filling system at a 3-bar pressure [53].

The marine substation is completed by an auxiliary system, a resonant circuit, and a measurement station. The main function of the auxiliary system is to distribute the power generated by the WEC units among the control systems and the circuit breakers. This installation is fed by the WEC generators, or it can be fed by the off-shore system when feeding by the WEC generators is not possible. The resonant circuit enables increased production by the installation. Lastly, the measurement station is supplied with different loads so that the LRS performance and its remote control can be widely studied.

This entire system has been subjected to several control mechanisms, such as Programmable Automation Controllers (PACs) for both the marine substation and the power inverter, as well as a Data Acquisition System, which was implemented for gathering electrical outputs from each wave converter [59].

6. Lysekil Research Site Modelling for Power Output Smoothen

6.1. LRS Input Data

The data used in this research were gathered from three linear generators, and each is attached to a cylindrical buoy (L1, L2, and L3) [51]. The voltage and current measurements from the three linear generators were sampled at 256 frames per second (921,600 frames in total). The data used were gathered from Uppsala University on 20 July 2009, at 22:00 h [60]. A boxplot analysis was carried out to determine the trends in each phase of the three generators, as shown in Figure 12. Taking into account the number of outliers in the quartiles, it can be inferred that the wave energy converters were subjected to recurrent overloading conditions, and the influence of such conditions on the power output should not be dismissed [61,62]. It is important to point out that when there are dips in the power output, the translator needs to reach a high speed to overcome this condition and deliver energy to the system [61]. This point was validated by the results in Figure 13 for Generator 1; the sample time was 10 s.

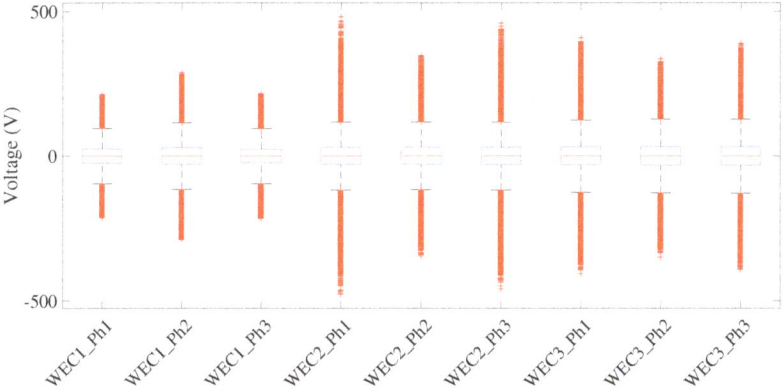

Figure 12. Boxplot analysis of each phase of the wave converters (WEC1, 2, and 3).

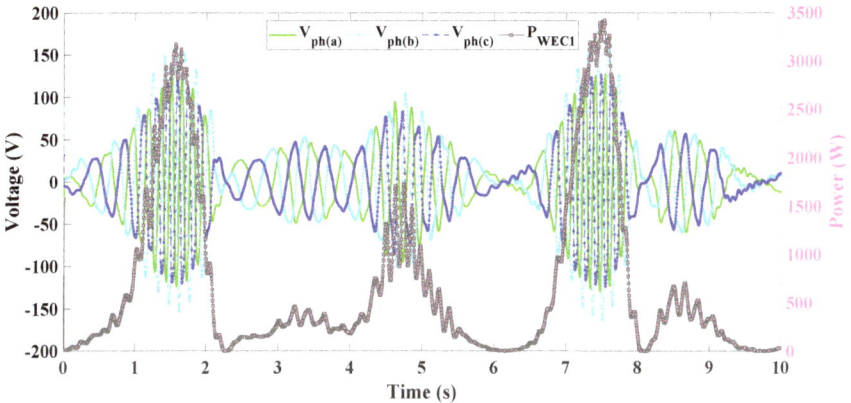

Figure 13. Voltages and power of $PWEC_1$.

At the Lysekil test site, several storage systems have been applied to reduce the power fluctuations from the wave generators, as fully explained in [21]. The alternative options for energy storage systems in the DC link include batteries and supercapacitors connected through a DC/DC converter. It was demonstrated in [21] that the installation of batteries and supercapacitors at the Lysekil test site can minimize the fluctuations in the power supplied to the grid. However, both storage systems present some limitations. The performance of batteries degrades over time, causing increased maintenance and increased costs. Supercapacitors are not suitable for power smoothing because the wave period is on the order of 10 s, necessitating over 5000 charge/discharge cycles per day, which is unattainable for supercapacitors since they have a much lower energy density. For this reason, in this study, the application of a Flywheel ESS was tested at the Lysekil test site in order to smooth the power output without offering mechanical degradation, even with multiple charge/discharge cycles.

6.2. LRS Modelling

6.2.1. General Description

The general structure of the Lysekil Wave Farm and the proposed wave power smoothing system is shown in Figure 14, which includes a wave farm, a filter, a Kinetic Energy Storage System (KESS), a DC/AC converter, and a marine transformer. Each block in the wave farm is described below:

1. The wave power farm consists of three WECs. For wave power output smoothing, the WECs were modeled with the measurement data collected from the three WECs located at the Lysekil Test Site (see Figure 14, WEC's Block) [56].
2. Filter: the purpose of the filter is to create a reference signal for the FESS in order to store wave energy in this device. To accomplish that, the filter decomposes the input signal (power from WECs) into two components: one that corresponds to the smoothed power output and another that corresponds to the fluctuating part of the signal and that must be stored in the FESS (Figure 14, Reference signal Block).
3. The KESS is formed by a high-speed flywheel that is connected to the DC link (see Figure 14, FESS Block).
4. The DC/AC inverter is an IGBT inverter that improves the wave farm power output by filling the power valleys with the energy stored in the KESS (see Figure 14, Grid-Side Converter Block).
5. Marine transformers are used for integrating the wave power output to the on-shore power grid. The transmission network at the Lysekil emplacement consists of two power transformers with five winding taps connected at each extreme of a submarine transmission line. In this study, only the first transformer was considered [63].

The aforementioned elements can function collectively to feed an isolated or grid-connected system.

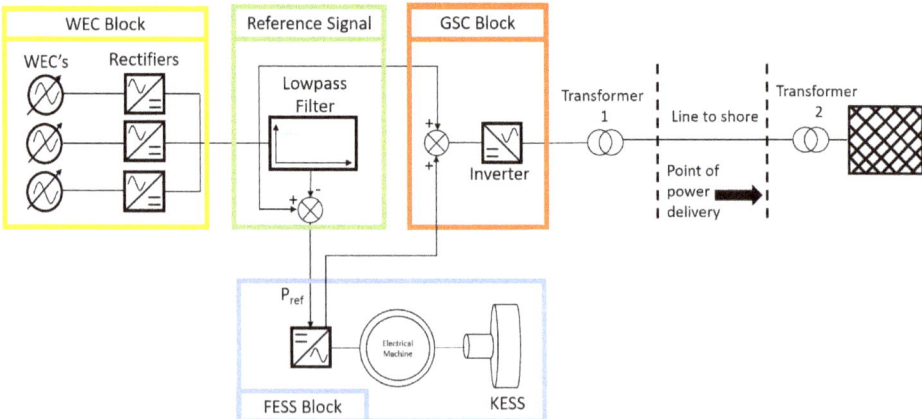

Figure 14. Circuit diagram applied of a linear wave energy system located in Lysekil Test Site.

6.2.2. WEC Units

Real power measurements of the three WEC units are used to demonstrate the KESS's ability to smooth the wave power output. Figure 15a) shows the sea state in terms of the buoy's wave elevation during the first half-hour of the study [60], and Figure 15b) represents the WEC power output registered during the first half-hour of the study. As can be seen in Figure 15a), the hour of study corresponds to a calm sea with a wave height that ranges between 27 and −22 cm, and Figure 15b) demonstrates that the medium power output is 2.55 kW.

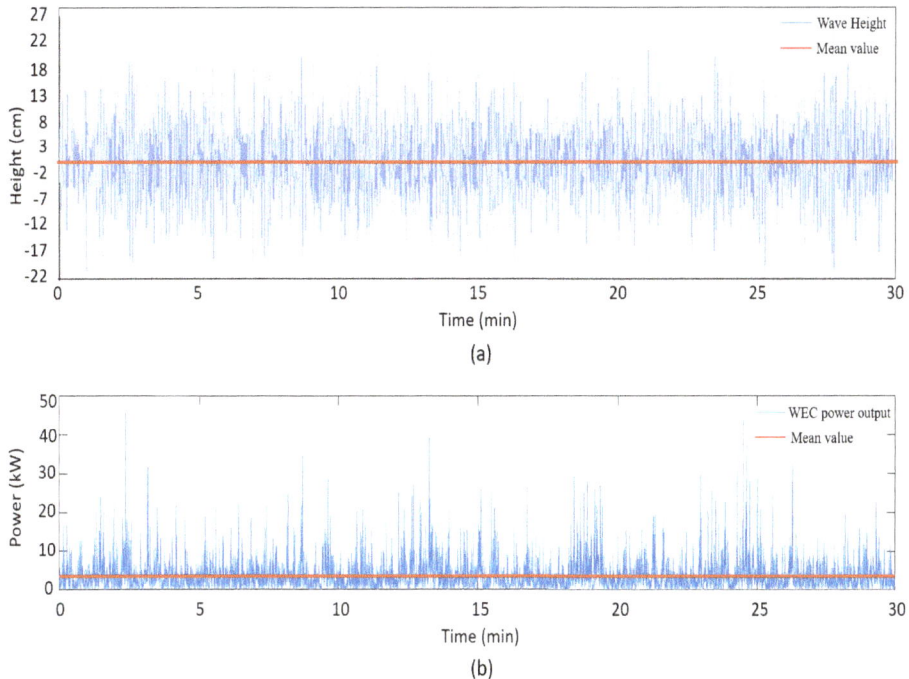

Figure 15. Buoy wave elevation (**a**) and WECs' power output (**b**) for the Lysekil wave farm.

6.2.3. Flywheel Model

In this study, the design data of a flywheel prototype, developed by Uppsala University, were used [64]. This prototype comprises a hollow cylinder, radial magnetic bearings, and an electrical machine that is capable of acting as either a motor or a generator. The kinetic energy that the FESS delivers is governed by (1), and it has an established maximum discharge depth of 25% with respect to the FESS' storage capacity.

To minimize the device's losses, two aspects were considered during the development of the model:

(a) First, the flywheel rotor is located inside a chamber, and it is suspended by magnetic bearings that allows them to work at quasi-vacuum conditions. Therefore standby losses are minimized [65].
(b) Secondly, the PMSM that is used as a motor/generator is composed of a coreless stator [66].

Finally, losses can be grouped and categorized as drag , electrical machine , and magnetic bearings losses.

- Drag losses occur when the FESS model is operated between 15 and 30 krpm. In this speed range, the air in the flywheel chamber develops a turbulent velocity regime. Under this condition, the power that dissipates because of the drag losses from a rotating cylinder in a concentrically cylindrical case can be expressed as (6) [67]:

$$P_d = \omega R F_d \tag{6}$$

where R is the cylinder radius, and F_d is the frictional force on the cylinder. The force induced by the viscous drag is expressed as (7):

$$F_d = \lambda \rho \pi R^3 L \omega^2 \tag{7}$$

where λ is the friction factor or drag coefficient obtained experimentally at low speed [64], ρ is the air density [64], and L is the cylinder length.

- For electrical machine losses in the FESS model, an electrical machine whose stator is coreless is considered; thus, the main source of losses in the machine is Joule losses, as expressed in (8):

$$P_i = RI^2 \tag{8}$$

- Lastly, the magnetic bearing losses associated with the magnetic bearing that is responsible for sustaining the flywheel rotor can be classified as hysteresis losses or eddy losses. At high flywheel speeds, the eddy losses are those that surpass the hysteresis losses [64]; thus, for the developed FESS model, magnetic bearing losses are calculated according to (9), and these losses are directly proportional to the square of the flywheel rotational speed.

$$P_{eddy} = K_{eddy} \hat{B}^2 f^2 \tag{9}$$

where K_{eddy} is a constant that depends on both the geometry and the material used in the magnetic bearing's design, $\hat{B}$ is the maximum magnetic flux, and f is the flywheel frequency.

6.2.4. Grid Connection

Both the KESS and WECs are connected to the grid through a grid-side converter (GSC) that is rated to the nominal power of the KESS and WECs at the farm. The GSC is assumed to have a constant efficiency of 98.5% (see Figure 14). The 2.9 km sea transmission line that connects the offshore and onshore substations is a pi model with a line resistance of 0.21 Ω/km [51].

7. LRS Power Output Smoothening

The power output smoothing process presented in this paper has three main steps. In the first step, the power output of several WECs is clustered at the wave farm connection point. Next, a filter is applied to the output signal of the clustering process in order to determine the power reference signal that must be stored in the flywheel. In the final step, a process of filling power valleys is implemented in order to achieve the reference power output signal given by a grid operator.

Figure 16 shows the complete block diagram of the proposed power output smoothing process. The experimental Lysekil Marine Platform in Sweden that is analyzed in this paper is formed by a permanent magnet linear generator WEC. Data used in this analysis come from a passive rectifier connected to a load. The total power of the wave farm is equal to the sum of the power injected by each individual WEC. Because there is no WEC regulation, the power output is highly stochastic, and this necessitates the use of the FESS system for power smoothing purposes. The flywheel control is only coordinated with the filtering stage. The PI controller has been designed using the pole placement technique tuned for the nominal values and results in Kp = 0.212 and Ki = 3.6. The yellow box corresponds to the first step, the green box corresponds to the second step, and the blue box corresponds to the last step of the developed power smoothing process.

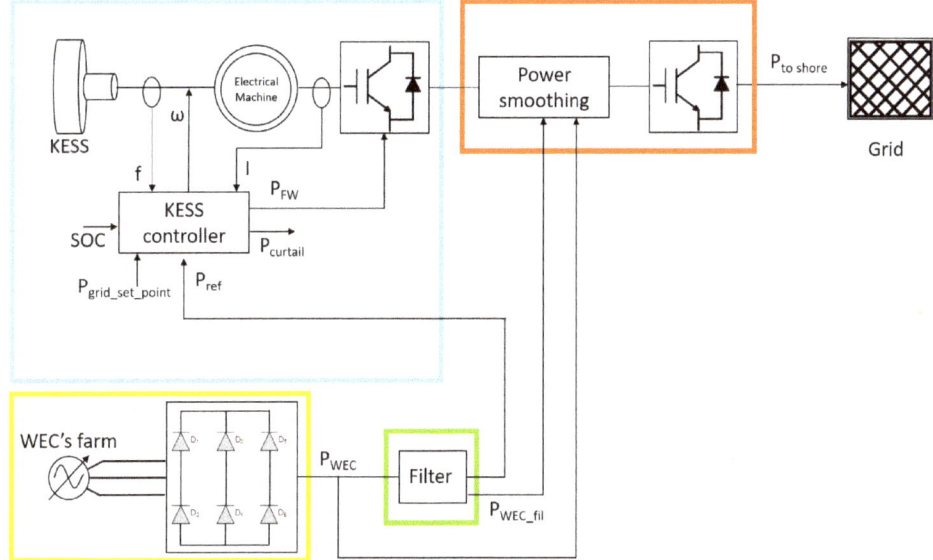

Figure 16. Schematic representation of the power output smoothening process.

7.1. First Stage: Generators Clustering

In this stage, the effect of clustering generators was analysed using several parameters. To do this, each generator (WEC1, WEC2, and WEC3) and its delivered wave active power were studied separately. Then, every two generators are clustered. Finally, the three converters were aggregated to a common node. The Lysekil WEC model is fully explained in [19], and it is composed of three WEC units, each of which is formed by a surface buoy connected to the permanent magnet linear generator moored to a foundation on the seabed [68]. The main parameters associated with the model are specified in Appendix.

The effect of the wave active power aggregation is summarized in Table 1, and it was examined in the context of the key performance indicators (KPIs), which are described below.

1. Peak Power, Average Power, and Peak/Average Ratio.

    As the number of generators clustered at the grid connection point increases, both the peak and average power values of the power output from the wave energy plant increase, leading to a simultaneous decrease in the peak/average power ratio.

2. Normalized Standard Deviation (nSD).

    The normalized standard deviation is measured in p.u. and gives a measure of the relative power output fluctuations. This value is more intuitive than the value of absolute fluctuations.

3. Point of delivery KPIs.

    Table 1 shows the power delivered to the grid by the wave energy farm at the grid connection point. From the results, it can be concluded that clustering the WEC units improves the power output in terms of the peak/average ratio and standard deviation. Nevertheless, the power output variability is not completely improved. Consequently, additional techniques need to be carried out in order to achieve this goal; these techniques include filtering and valley filling, and their procedures are fully shown in Table 1.

Table 1. Wave power data analysis from the three generators and the effect of aggregation.

	Peak P_{shore} (kW)	Mean P_{shore} (kW)	Peak/Ave. P_{shore}	Normalized Standard Deviation	Energy to Shore (kWh)	Grid Losses (Wh)
WEC1	9.94	0.81	12.28	1.54	0.81	0.85
WEC2	40.82	1.29	32.71	1.75	1.25	2.42
WEC3	30.09	1.35	22.32	1.52	1.35	2.48
WEC (1 + 2)	42.83	2.06	20.82	1.23	2.05	4.12
WEC (1 + 3)	33.98	2.16	15.75	1.12	2.15	4.26
WEC (2 + 3)	45.30	2.60	17.45	1.16	2.59	6.20
All WECs	45.58	3.41	13.39	0.96	3.4	8.840

7.2. Second Stage: Filtering and KESS Stage

To further improve the power output signal from the wave farm at Lysekil, a low-pass filter was designed. The filter has minimal complexity for its implementation in the real installation. The proposed filter is composed of a low-pass filter with a zero-order hold block. The combined filter is governed by (10), where f_{fil} is the filter's frequency. Figure 17 shows the block diagram of the filter control step.

$$T = \frac{1}{2\pi f_{fil} s + 1} \quad (10)$$

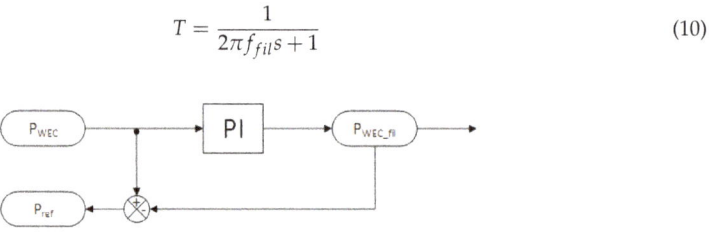

Figure 17. Block diagram of the filtering process.

Wave energy is characterized by highly stochastic behavior, including variable and random wave energy frequency, amplitude, and direction. At the Lysekil wave power plant, there are no generator controllers, and consequently, the high variability in wave natural energy resources translates into high variability in the power that is injected into the grid, as reported in [19,57]. Figure 18 shows the power output from the linear generators and the filtered power in the intermediate filtering stage. Note that the filtered signal corresponds to an intermediate stage before the KESS process since the flywheel is connected to the DC bus and the harmonic content at this stage is not an issue for the grid because it is located after the inverter.

As previously mentioned, the wave power plant offers large oscillations for short time periods. For this reason, the storage system to be used should meet certain criteria, such as quick charge/discharge cycles with no electrochemical degradation, a high power density, and a high capability of meeting stipulated requirements [69].

In this research, high-speed buffering was applied to the wave power prototype located at the Lysekil Test Site. The Lysekil flywheel was modeled as a hollow cylinder, as presented in [42,64]. The size of the flywheel can be selected according to the energy required on the basis of minimum and maximum thresholds.

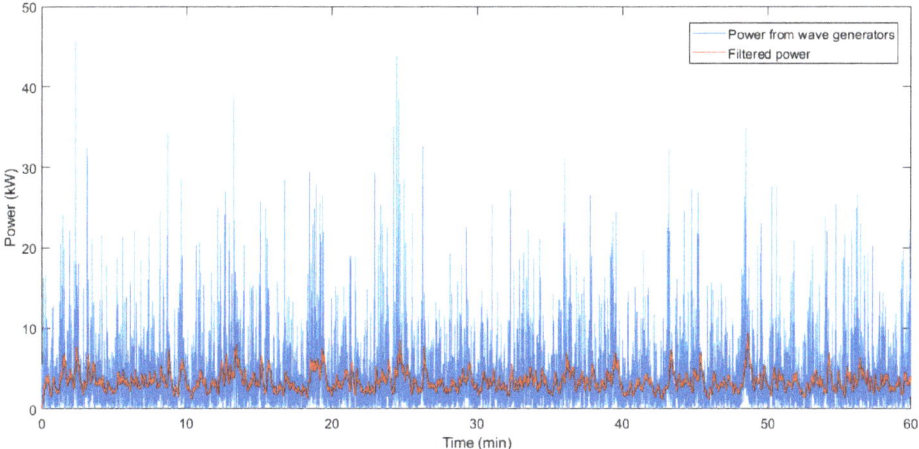

Figure 18. Results of the filtering control system over the wave power output.

Figure 19 shows the block diagram of the flywheel energy storage control, and the charge and discharge control strategy of the KESS is shown in Figure 20. A reference signal (P_{ref-FW}) is used to control the energy storage of the flywheel. The difference between the output power of the WEC array ($P_{WEC_{farm}}$) and the filtered power (P_{fil}) is used as the power reference signal (P_{ref-FW}) of the flywheel. If $P_{WEC_{farm}}$ is larger than P_{fil} ($P_{ref-FW} = P_{WEC_{farm}} - P_{fil} > 0$), the difference between both signals is sent as P_{ref-FW} to the KESS, and the flywheel stores energy. Once the KESS achieves its maximum state of charge (SOC) (0.9 p.u.), the excess power is curtailed ($P_{curt} = P_{ref-FW} - P_{nom_{FW}}$). If $P_{ref-FW} = P_{WEC_{farm}} - P_{fil} < 0$, then the KESS is in discharge mode and the power injected into the system is equal to P_{ref-FW}. The KESS discharges energy until the SOC_{min} limit is achieved. In this study, the SOC_{min} = 0.25 p.u. If SOC_{min} is reached, there is no possibility for the KESS to inject energy into the system, so the P_{WEC} is sent to the shore. This control strategy will be linked with the wave generator controller to smooth the wave power farm output.

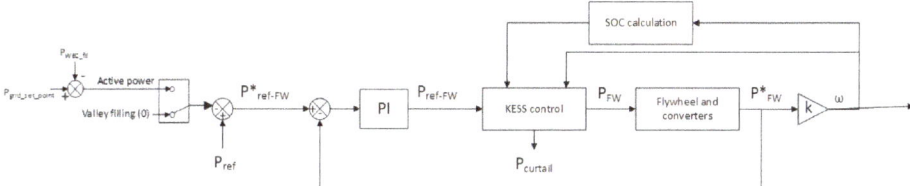

Figure 19. Block diagram of the FESS control process.

Figures 21 and 22 show an example of the control strategy in two different periods of time. The lines in Figure 21 (i.e., A, B, and C) and the lines in Figure 22 (i.e., D, E, and F) represent the time intervals in which the flywheel performance experiences a change. The space between lines A and B corresponds to the KESS charge mode, in which the reference signals are positive (P_{ref-FW}, orange line). Once the KESS achieves the maximum SOC (SOC_{max}=0.9*SOC_{nom}) at line B, the power is then curtailed, and the power to shore (P_{out}, blue line) remains constant with the energy fed by the KESS. At point C, the reference signal becomes negative, and the KESS system is in discharge mode. Figure 22 shows a period of time in which the KESS charge and discharge mode is around the minimum KESS SOC. At the initial instance portrayed in Figure 22, the FESS SOC is situated at its minimum value (SOC_{min}=0.25*SOC_{nom}). In the first instance of the simulation (i.e., up to line D), the installation's power output (P_{out}, blue line) roughly follows the output reference signal of the wave emplacement once it has been filtered. This is because the KESS experiences constant charge/discharge cycles around its SOC_{min}, and this is caused by the zero-crossing oscillations associated with the KESS' reference signal. The KESS is charged during the time interval that ranges from line D to line E because of the positive reference signal, which allows the wave installation to maintain a constant power output of 3 kW until the time at which line F is reached. From this point, it is observed that the KESS does not possess enough stored energy to keep the installation's power output at a constant value, so it tracks back the filtered signal of the power output ($P_{from farm}$, green line).

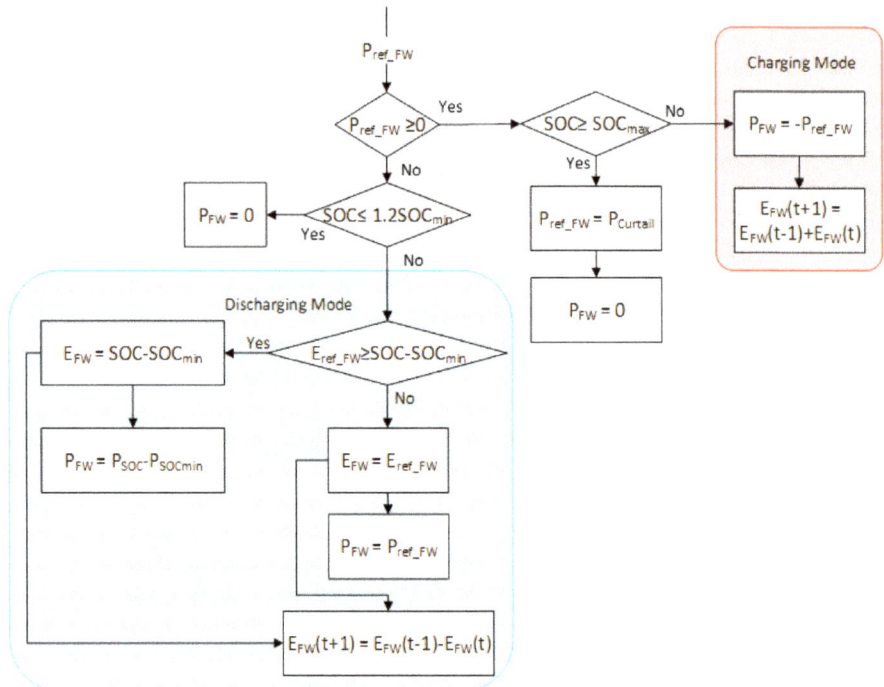

Figure 20. KESS control flowchart.

Several simulations were performed (see Figure 23), and these simulations involved a filtered frequency range between 256 and 2048 Hz and a flywheel with a moment of inertia varying between I_{nom} and $5*I_{nom}$. It must be noted that the flywheel control is linked to the filter control. Consequently, the optimal flywheel size depends not only on the inverter capacity but also on the cut-off frequency filter. Therefore, it can be deduced that a filter with a cut-off frequency of 256 Hz is the frequency that offers the minimum curtailment, the maximum energy to the shore, and greater mean power to the shore.

Energies **2019**, 12, 2196

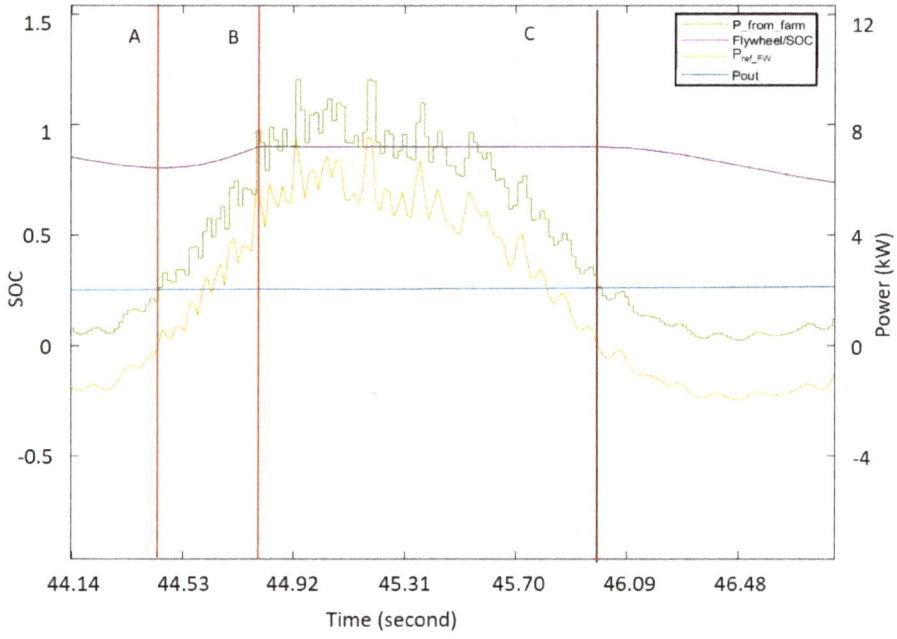

Figure 21. Example of the KESS Control's performance I.

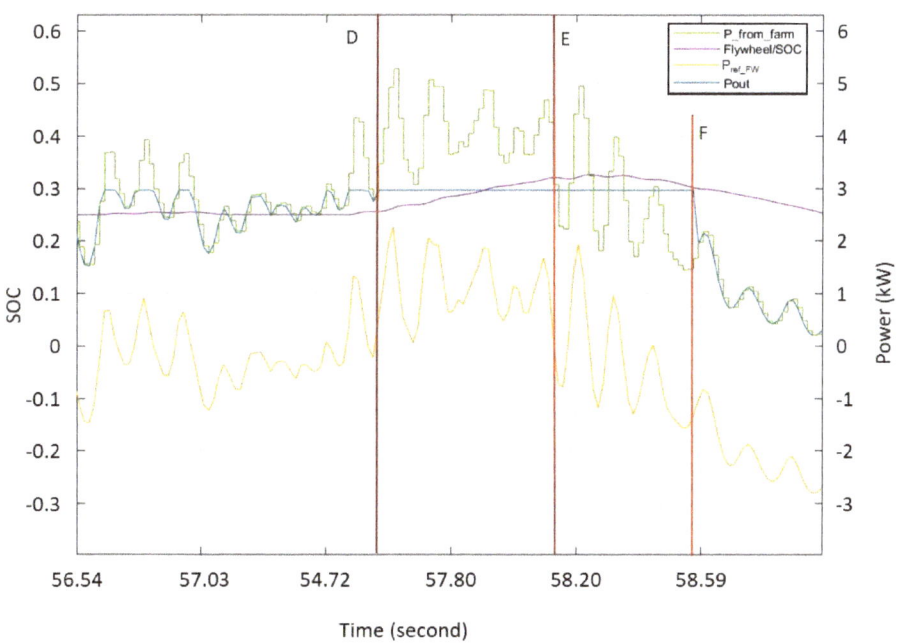

Figure 22. Example of the KESS Control's performance II.

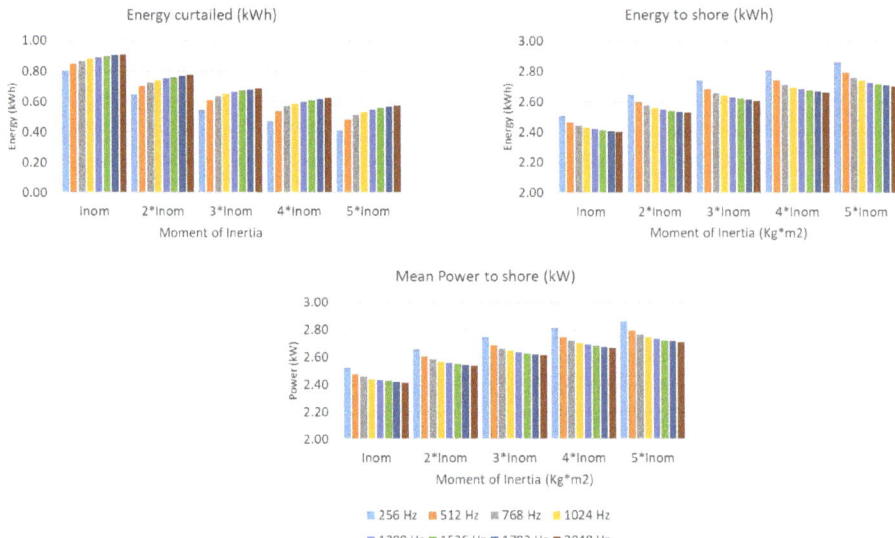

Figure 23. Comparison of energy curtailed, energy to shore, and mean power to shore between different filtered frequencies and flywheel moments of inertia.

Once the optimal filter frequency for this installation was selected ($f_{fil_{opt}}$ = 256 Hz), a new experiment with a variable moment of inertia between 1 and 8 times the nominal inertia moment was performed. The objective of this experiment was to determine the optimal size of the flywheel.

Figure 24 shows this experiment's results, which illustrate the relative improvement in the three KPIs. It can be seen that the improvement in the energy curtailed ranges from 20% (I_{fw} = 2*I_{nom}) to 11% (for I_{fw} higher than 5*I_{nom}) compared with the I_{nom} value. The improvement in the energy to the shore varies from 6% to 1%, and it is almost constant from I_{fw} = 2*I_{nom}. For the reduction in energy losses, the more drastic reduction is found between I_{fw} = 2*I_{nom} and I_{fw} = 4*I_{nom}, and from I_{fw} = 5*I_{nom}, the reduction improvement decreases slowly. It can be concluded that the best size of the flywheel for this application is the one with a moment of inertia of $I_{fw_{opt}}$ = 5*I_{nom}. For the optimal flywheel size, the cylinder height is calculated by (11), where h is the height in meters of the prototype located at the Lysekil Test Site [70,71]:

$$h_{FW_{opt}} = \frac{2 * E_{FW_{max}} * h_{ref}}{I_{FW_{opt}} * \omega^2} \quad (11)$$

The optimal cylinder height is $h_{FW_{opt}} = 0.9 * h$, where h_{ref} is the flywheel's height in the original prototype, $E_{FW_{max}}$ is the maximum energy delivered by the flywheel, and $I_{FW_{opt}}$ is the optimal moment of inertia of the device. The operating speed ranges from ω_{min} = 15 krpm to ω_{max} = 30 krpm, with a minimum and maximum SOC of SOC_{min} = 25% and SOC_{max} = 90%, respectively.

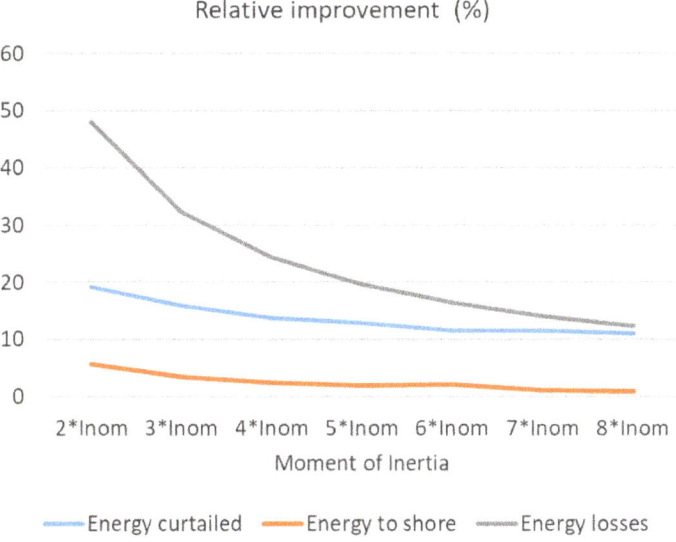

Figure 24. Relative improvements in KPIs.

7.3. Active Power Control

The above sections describe both the filters and flywheel that were used to improve the LRS's output signal. This subsection analyzes the capacity of the proposed hybrid wave energy farm's flywheel system for tracking a certain power set-point given by the grid operator. The blue curve in Figure 25 shows the WEC output power delivered to the grid, the red curve shows the power delivered by the grid-connected wave-flywheel, and the green curve represents the flywheel's SOC. It can be seen that the hybrid wave-flywheel system can deliver the 2 kW set-point established by the grid operator. However, during the time intervals in which the flywheel does not have enough stored energy, the hybrid installation is not capable of tracking the set-point established by the grid, and at this point, there is a need to regulate the energy stored in the flywheel.

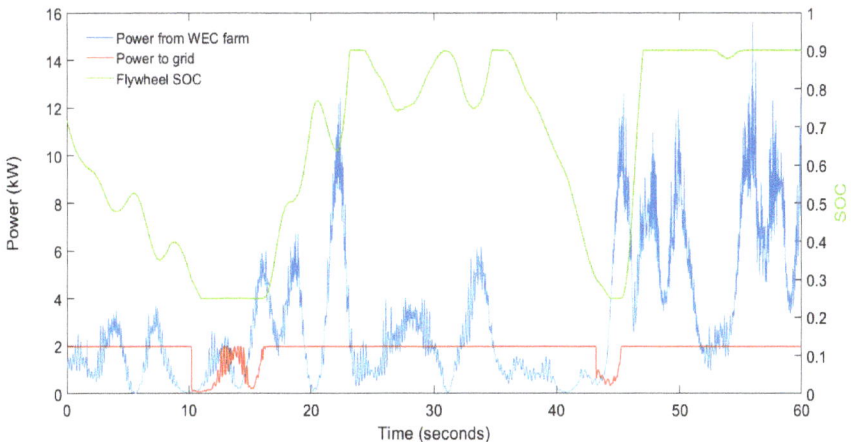

Figure 25. Power-to-shore following a grid set-point.

8. Discussion

Table 2 summarizes the KPIs associated with the wave power smoothing control proposed in this paper. In analyzing Tables 1 and 2, the following conclusions can be drawn:

Table 2. Summarized results for different smoothening stages.

	Peak P_{shore} (kW)	Mean P_{shore} (kW)	Peak/Ave. P_{shore}	Grid Losses (Wh)	Curtailed Energy (kWh)	Energy to Shore (kWh)	Normalized Standard Deviation
WEC clustering	45.58	3.41	13.39	8.84	–	3.40	0.96
LPF at 256 Hz	9.23	2.52	3.66	12.72	–	2.51	0.58
KESS (3.18 kg.m^2)	9.23	2.91	3.17	4.37	0.36	2.92	0.49
Active power control	4.56	2.68	1.71	3.47	0.70	2.68	0.44

1. First, aggregating a group of generators into one cluster results in a decreased Peak/Average power ratio and an improved normalized standard deviation (nSD).
2. In addition, two clustered wave generators share the same magnitude order when referencing the peak values; this results in a notable increase in both the energy losses and the energy delivered to the grid (to the shore). This can be seen in Table 1, in which the results of aggregating Generators 2 and 3 are reported.
3. Furthermore, the cut-off frequency of the low-pass filter is critical for the smoothing process. The study found an optimal cut-off frequency of 256 Hz; it not only filters the output power but also reduces the energy curtailment. Increasing the cut-off frequency of the filter above this optimal value does not decrease the losses in the substation or flywheel, but it increases the energy curtailment, which is not efficient.
4. Moreover, increasing the moment of inertia of the flywheel (that is, increasing the cylinder height) results in a considerable decrease in the energy curtailment. Nevertheless, an appropriate criterion for choosing the cylinder size could be based on the related improvement in (1) the energy that is delivered to the shore and (2) the curtailed energy. This optimal value is achieved for a moment of inertia that is equal to 3.18 kg m^2, which corresponds to a flywheel size that is roughly 10% of the inverter's size.
5. In conclusion, almost all of the analyzed KPIs are considerably improved if the aforementioned smoothing techniques are carried out, such as the peak power to the shore, peak/mean power to the shore ratio, normalized standard deviation, and grid connection losses, all of which are decreased. These findings are a clear indicator of the effectiveness of the proposed simulated approach.

9. Conclusions

Flywheels are found to be a good option for the smoothing process. It was confirmed that utilizing flywheel systems coupled to wave farms improves the power output, and the grid integration of this technology is improved accordingly.

In this paper, a hybrid wave/KESS installation control is proposed. This control combines wave energy generator clustering and filtering techniques with the control of a flywheel in order to improve the power output of the hybrid installation. The proposed control was applied at the Lysekil Wave Energy Site located in Sweden by using real measurements gathered from the wave energy generators. It was established that the proposed control smooths the power output delivered by the marine energy installation to the grid and allows the connection of the wave/FESS installation to power grids, even weak ones. Moreover, it was shown that the hybrid plant can follow a power reference signal imposed by the grid operator. Consequently, the proposed combination of the wave energy farm control and flywheel energy storage system can be grid-connected as a controllable distributed energy resource.

Therefore, the hybrid system is capable of offering enough flexibility to the distributed system operator in terms of ancillary services.

Author Contributions: All authors contributed to the design and implementation of the research, to the analysis of the results and to the writing of the manuscript.

Funding: This research received no external funding

Conflicts of Interest: The authors declare no conflict of interest.

Appendix

WEC main parameters

Nominal power at 0.7 m/s	10 kW
Voltage (line to line, rms at 0.7 m/s), V_d	200 V
Generator resistance, R_G	$0.44 \pm 1\%$ Omega
Generator inductance, L_S	11.7 mH
Air gap	3 mm
Size of the magnet block	$6.5 \times 35 \times 100$ mm^3
Pole width, w_p	50 mm
Number of stator sides	4
Vertical stator length	1264 mm
Vertical translator length	1867 mm
Transator resp. stator width	400 mm
Translator weight	1000 kg

References

1. Organization for Economic Cooperation and Development (OECD). *World Energy Outlook*; Technical Report; International Energy Agency: Paris, France, 2016; ISBN: 978-92-64-26495-3.
2. Uihlein, A.; Magagna, D. Wave and tidal current energy—A review of the current state of research beyond technology. *Renew. Sustain. Energy Rev.* **2016**, *58*, 1070–1081. [CrossRef]
3. Falnes, J. A review of wave-energy extraction. *Mar. Struct.* **2007**, *20*, 185–201. [CrossRef]
4. Prakash, S.; Mamun, K.; Islam, F.; Mudliar, R.; Pau'u, C.; Kolivuso, M.; Cadralala, S. Wave Energy Converter: A Review of Wave Energy Conversion Technology. In Proceedings of the 3rd Asia-Pacific World Congress on Computer Science and Engineering, Nadi, Fiji, 5–6 December 2016; pp. 71–77.
5. Marine Energy—Wave Device The European Marine Energy Centre. 2019. Available online: http://www.emec.org.uk/marine-energy/wave-devices/ (accessed on 24 February 2019).
6. Hong, Y.; Waters, R.; Boström, C.; Eriksson, M.; Engström, J.; Leijon, M. Review on electrical control strategies for wave energy converting systems. *Renew. Sustain. Energy Rev.* **2014**, *31*, 329–342. [CrossRef]
7. Leijon, J. Simulation of a Linear Wave Energy Converter with Different Damping Control Strategies for Improved Wave Energy Extraction. Master's Thesis, Uppsala University, Uppsala, Sweden, 2016.
8. Babarit, A.; Guglielmi, M.; Clément, A.H. Declutching control of a wave energy converter. *Ocean Eng.* **2009**, *36*, 1015–1024. [CrossRef]
9. Brito, M.; Teixeira, L.; Canelas, R.B.; Ferreira, R.M.; Neves, M.G. Experimental and Numerical Studies of Dynamic Behaviors of a Hydraulic Power Take-Off Cylinder Using Spectral Representation Method. *ASME J. Tribol.* **2017**, *140*, 021102. [CrossRef]
10. De O. Falcão, A.F. Modelling and control of oscillating-body wave energy converters with hydraulic power take-off and gas accumulator. *Ocean Eng.* **2007**, *34*, 2021–2032.
11. De O. Falcão, A.F. Phase control through load control of oscillating-body wave energy converters with hydraulic PTO system. *Ocean Eng.* **2008**, *35*, 358–366.
12. Howlader, A.M.; Urasaki, N.; Yona, A.; Senjyu, T.; Saber, A.Y. A review of output power smoothing methods for wind energy conversion systems. *Renew. Sustain. Energy Rev.* **2013**, *26*, 135–146. [CrossRef]
13. Rahm, M.; Svensson, O.; Boström, C.; Waters, R.; Leijon, M. Experimental results from the operation of aggregated wave energy converters. *IET Renew. Power Gener.* **2012**, *6*, 149–160. [CrossRef]

14. Venugopal, V.; Smith, G.H. The effect of wave period filtering on wave power extraction and device tuning. *Ocean Eng.* **2007**, *34*, 1120–1137. [CrossRef]
15. Godoy-Diana, R.; Czitrom, S.P. On the tuning of a wave-energy driven oscillating -water-column seawater pump to polychromatic waves. *Ocean Eng.* **2007**, *34*, 2374–2384. [CrossRef]
16. Bedard, R.; Hagerman, G. *E2I EPRI Assessment Offshore Wave Energy Conversion Devices*; Electrical Innovation Institute: Washington, DC, USA, 2004.
17. Moreno-Torres, P.; Blanco, M.; Navarro, G.; Lafoz, M. Power smoothing system for wave energy converters by means of a supercapacitor-based energy storage system. In Proceedings of the 2015 17th European Conference on Power Electronics and Applications (EPE'15 ECCE-Europe), Geneva, Switzerland, 8–10 September 2015; pp. 1–9.
18. Blavette., A.; O'Sullivan, D.L.; Lewis, T.; Egan, M.G. Grid Integration of Wave and Tidal Energy. In Proceedings of the International Conference on Offshore Mechanics and Arctic Engineering, Rotterdam, The Netherlands, 19–24 June 2011; pp. 749–758.
19. Lejerskog, E.; Boström, C.; Hai, L.; Waters, R.; Leijon, M. Experimental results on power absorption from a wave energy converter at the Lysekil wave energy research site. *Renew. Energy* **2015**, *77*, 9–14. [CrossRef]
20. Ma, T.; Yang, H.; Lu, L. Development of hybrid battery–supercapacitor energy storage for remote area renewable energy systems. *Appl. Energy* **2015**, *153*, 56–62. [CrossRef]
21. Parwal, A.; Fregelius, M.; Temiz, I.; Göteman, M.; de Oliveira, J.G.; Boström, C.; Leijon, M. Energy management for a grid-connected wave energy park through a hybrid energy storage system. *Appl. Energy* **2018**, *231*, 399–411. [CrossRef]
22. Krings, A.; Soulard, J. Overview and comparison of iron loss models for electrical machines. *J. Electr. Eng.* **2010**, *10*, 162–169.
23. Barranger, J. *Hysteresis and Eddy-Current Losses of a Transformer Lamination Viewed as an Application of the Poynting Theorem*; NASA Technical Note (TN-D-3114); National Aeronautics and Space Administration: Washington, DC, USA, 1965.
24. Henderson, R. Design, simulation, and testing of a novel hydraulic power take-off system for the Pelamis wave energy converter. *Renew. Energy* **2006**, *31*, 271–283. [CrossRef]
25. Zhao, X.; Yan, Z.; Zhang, X.P. A wind-wave farm system with self-energy storage and smoothed power output. *IEEE Access* **2016**, *4*, 8634–8642. [CrossRef]
26. Maslen, E.H.; Schweitzer, G. *Magnetic Bearings: Theory, Design, and Application to Rotating Machinery*; Springer: Berlin, Germany, 2009.
27. Bolund, B.; Bernhoff, H.; Leijon, M. Flywheel energy and power storage systems. *Renew. Sustain. Energy Rev.* **2007**, *11*, 235–258. [CrossRef]
28. Arani, A.K.; Karami, H.; Gharehpetian, G.; Hejazi, M. Review of Flywheel Energy Storage Systems structures and applications in power systems and microgrids. *Renew. Sustain. Energy Rev.* **2017**, *69*, 9–18. [CrossRef]
29. Infield, D.; Hill, J. *Literature Review: Electrical Energy Storage for Scotland*; University of Strathclyde: Glasgow, UK, 2015.
30. Amaris, H.; Alonso, M.; Alvarez, C. *Reactive Power Management of Power Networks with Wind Generation*; Springer: Berlin, Germany, 2013.
31. Amaris, H.; Alonso, M. Coordinated reactive power management in power networks with wind turbines and FACTS devices. *Energy Convers. Manag.* **2011**, *52*, 2575–2586. [CrossRef]
32. Asociación Española de Normalización. *UNE-EN 50160: Voltage Characteristics of Electricity Supplied by Public Electricity Networks*; Technical Report; Asociación Española de Normalización: Madrid, Spain, 2011; 2011/A1:2015.
33. Amiryar, M.E.; Pullen, K.R. A review of flywheel energy storage system technologies and their applications. *Appl. Sci.* **2017**, *7*, 286. [CrossRef]
34. Diaz-Gonzalez, F.; Bianchi, F.D.; Sumper, A.; Gomis-Bellmunt, O. Control of a flywheel energy storage system for power smoothing in wind power plants. *IEEE Trans. Energy Convers.* **2014**, *29*, 204–214. [CrossRef]
35. Gurumurthy, S.R.; Agarwal, V.; Sharma, A. Optimal energy harvesting from a high-speed brushless DC generator-based flywheel energy storage system. *IET Electr. Power Appl.* **2013**, *7*, 693–700. [CrossRef]
36. Wang, L.; Yu, J.Y.; Chen, Y.T. Dynamic stability improvement of an integrated offshore wind and marine-current farm using a flywheel energy-storage system. *IET Renew. Power Gener.* **2011**, *5*, 387–396. [CrossRef]

37. Park, J.D.; Kalev, C.; Hofmann, H.F. Control of high-speed solid-rotor synchronous reluctance motor/generator for flywheel-based uninterruptible power supplies. *IEEE Trans. Ind. Electron.* **2008**, *55*, 3038–3046. [CrossRef]
38. Li, W.; Chau, K.; Ching, T.; Wang, Y.; Chen, M. Design of a high-speed superconducting bearingless machine for flywheel energy storage systems. *IEEE Trans. Appl. Supercond.* **2015**, *25*. [CrossRef]
39. Sihler, C.; Miri, A.M. A stabilizer for oscillating torques in synchronous machines. *IEEE Trans. Ind. Appl.* **2005**, *41*, 748–755. [CrossRef]
40. Recheis, M.N.; Schweighofer, B.; Fulmek, P.; Wegleiter, H. Selection of magnetic materials for bearingless high-speed mobile flywheel energy storage systems. *IEEE Trans. Magn.* **2014**, *50*, 1–4. [CrossRef]
41. Faraji, F.; Majazi, A.; Al-Haddad, K. A comprehensive review of Flywheel Energy Storage System technology. *Renew. Sustain. Energy Rev.* **2017**, *67*, 477–490.
42. Abrahamsson, J.; Hedlund, M.; Kamf, T.; Bernhoff, H. High-speed kinetic energy buffer: Optimization of composite shell and magnetic bearings. *IEEE Trans. Ind. Electron.* **2014**, *61*, 3012–3021. [CrossRef]
43. Ibarra-Berastegi, G.; Sáenz, J.; Ulazia, A.; Serras, P.; Esnaola, G.; Garcia-Soto, C. Electricity production, capacity factor, and plant efficiency index at the Mutriku wave farm (2014–2016). *Ocean Eng.* **2018**, *147*, 20–29. [CrossRef]
44. Tedeschi, E.; Santos-Mugica, M. Modeling and control of a wave energy farm including energy storage for power quality enhancement: The bimep case study. *IEEE Trans. Power Syst.* **2014**, *29*, 1489–1497. [CrossRef]
45. California Energy Commission. *Flywheel Systems for Utility Scale Energy Storage: A Transformative Flywheel Project for Commercial Readiness*; Technical Report; California Energy Commission: Union City, CA, USA, 2019.
46. Zhou, Z.; Benbouzid, M.; Charpentier, J.F.; Scuiller, F.; Tang, T. A review of energy storage technologies for marine current energy systems. *Renew. Sustain. Energy Rev.* **2013**, *18*, 390–400. [CrossRef]
47. Hebner, R.; Beno, J.; Walls, A. Flywheel batteries come around again. *IEEE Spectr.* **2002**, *39*, 46–51. [CrossRef]
48. Bleuler, H.; Sandtner, J.; Regamey, Y.J.; F Barrot, A. Passive Magnetic Bearings for Flywheels. Solid Mechanics and Its Applications 2005. Available online: http://www.researchgate.net/publication/228554073_Passive_Magnetic_Bearings_for_Flywheels (accessed on 24 February 2019).
49. Wang, W.; Hofmann, H.; Bakis, C.E. Ultrahigh speed permanent magnet motor/generator for aerospace flywheel energy storage applications. In Proceedings of the IEEE International Conference on Electric Machines and Drives, San Antonio, TX, USA, 15 May 2005; pp. 1494–1500.
50. International Energy Agency. *Implementing Agreement on Ocean Energy Systems*; Technical Report; IEA-OES Executive Committee: Lisbon, Portugal, 2009.
51. Boström, C. Electrical Systems for Wave Energy Conversion. Ph.D. Thesis, Uppsala University, Uppsala, Sweden, 2011.
52. Waters, R.; Engström, J.; Isberg, J.; Leijon, M. Wave climate of the Swedish west coast. *Renew. Energy* **2009**, *34*, 1600–1606. [CrossRef]
53. Rahm, M.; Boström, C.; Svensson, O.; Grabbe, M.; Bulow, F.; Leijon, M. Offshore underwater substation for wave energy converter arrays. *IET Renew. Power Gener.* **2010**, *4*, 602–612. [CrossRef]
54. Rémouit, F.; Chatzigiannakou, M.A.; Bender, A.; Temiz, I.; Sundberg, J.; Engström, J. Deployment and Maintenance of Wave Energy Converters at the Lysekil Research Site: A Comparative Study on the Use of Divers and Remotely-Operated Vehicles. *J. Mar. Sci. Eng.* **2018**, *6*, 39. [CrossRef]
55. Melo, A.B.; Villate, J.L. *Annual Report 2016*; Technical Report; Executive Committee of Ocean Energy Systems: Lisbon, Portugal, 2016.
56. Lysekil Wave Energy Site. 2018. Available online: https://tethys.pnnl.gov/annex-iv-sites/lysekil-wave-energy-site (accessed on 6 May 2018).
57. Leijon, M.; Boström, C.; Danielsson, O.; Gustafsson, S.; Haikonen, K.; Langhamer, O.; Strömstedt, E.; Stålberg, M.; Sundberg, J.; Svensson, O.; et al. Wave energy from the North Sea: Experiences from the Lysekil research site. *Surv. Geophys.* **2008**, *29*, 221–240. [CrossRef]
58. Lejerskog, E.; Gravråkmo, H.; Savin, A.; Strömstedt, E.; Tyrberg, S.; Haikonen, K.; Krishna, R.; Boström, C.; Rahm, M.; Ekström, R.; et al. Lysekil research site, Sweden: A status update. In Proceedings of the 9th European Wave and Tidal Energy Conference, Southampton, UK, 5–9 September 2011.
59. Svensson, O.; Boström, C.; Rahm, M.; Leijon, M. Description of the control and measurement system used in the low voltage marine substation at the Lysekil research site. In Proceedings of the 8th European Wave and Tidal Energy Conference, Uppsala, Sweden, 7–10 September 2009.

60. Mätdata-Islandsberg. 2018. Available online: http://islandsberg.angstrom.uu.se/ (accessed on 22 April 2018).
61. Boström, C.; Waters, R.; Lejerskog, E.; Svensson, O.; Stalberg, M.; Stromstedt, E.; Leijon, M. Study of a wave energy converter connected to a nonlinear load. *IEEE J. Ocean. Eng.* **2009**, *34*, 123–127. [CrossRef]
62. Castellucci, V.; García-Terán, J.; Eriksson, M.; Padman, L.; Waters, R. Influence of sea state and tidal height on wave power absorption. *IEEE J. Ocean. Eng.* **2017**, *42*, 566–573. [CrossRef]
63. Waters, R.; Stålberg, M.; Danielsson, O.; Svensson, O.; Gustafsson, S.; Strömstedt, E.; Eriksson, M.; Sundberg, J.; Leijon, M. Experimental results from sea trials of an offshore wave energy system. *Appl. Phys. Lett.* **2007**, *90*, 034105. [CrossRef]
64. Hedlund, M. Electrified Integrated Kinetic Energy Storage. Ph.D. Thesis, Uppsala University, Uppsala, Sweden, 2017.
65. Hedlund, M.; Abrahamsson, J.; Pérez-Loya, J.J.; Lundin, J.; Bernhoff, H. Eddy currents in a passive magnetic axial thrust bearing for a flywheel energy storage system. *Int. J. Appl. Electromagn. Mech.* **2017**, *54*, 389–404. [CrossRef]
66. Hedlund, M.; Lundin, J.; de Santiago, J.; Abrahamsson, J.; Bernhoff, H. Flywheel energy storage for automotive applications. *Energies* **2015**, *8*, 10636–10663. [CrossRef]
67. Gorland, S.H.; Kempke, E.E., Jr.; Lumannick, S. *Experimental Windage Losses for Close Clearance Rotating Cylinders in the Turbulent Flow Regime*; Technical Report; NASA Technical Memorandum: Washington, DC, USA, 1970.
68. Bostrom, C.; Leijon, M. Operation analysis of a wave energy converter under different load conditions. *IET Renew. Power Gener.* **2011**, *5*, 245–250. [CrossRef]
69. Larsson, K. Investigation of a Wave Energy Converter with a Flywheel and a Corresponding Generator Design. Master's Thesis, Chalmers University of Technology, Göteborg, Sweden, 2012.
70. Abram, N.J.; McGregor, H.V.; Tierney, J.E.; Evans, M.N.; McKay, N.P.; Kaufman, D.S.; Thirumalai, K.; Martrat, B.; Goosse, H.; Phipps, S.J.; et al. Early onset of industrial-era warming across the oceans and continents. *Nature* **2016**, *536*, 411. [CrossRef]
71. Tornelli, C.; Zuelli, R.; Marinelli, M.; Rezkalla, M.M.; Heussen, K.; Morch, A.Z.; Cornez, L.; Zaher, A.S.; Catterson, V. *Key Requirements for Future Control Room Functionality*; Technical Report, The ELECTRA EU Research Project; University of Oldenburg: Oldenburg, Germany, 2015.

© 2019 by the authors. Licensee MDPI, Basel, Switzerland. This article is an open access article distributed under the terms and conditions of the Creative Commons Attribution (CC BY) license (http://creativecommons.org/licenses/by/4.0/).

Article

A Remotely Controlled Sea Level Compensation System for Wave Energy Converters

Mohd Nasir Ayob [1,2,*], Valeria Castellucci [1], Johan Abrahamsson [1] and Rafael Waters [1]

[1] Swedish Centre for Renewable Electric Energy Conversion, Division of Electricity, Department of Engineering Sciences, The Angstrom Laboratory, P.O Box 534, SE-75121 Uppsala, Sweden; valeria.castellucci@angstrom.uu.se (V.C.); johan.abrahamsson@angstrom.uu.se (J.A.); rafael.waters@angstrom.uu.se (R.W.)
[2] School of Mechatronic Engineering, Universiti Malaysia Perlis, Arau 02600, Perlis, Malaysia
* Correspondence: nasir.ayob@angstrom.uu.se; Tel.: +46-18-471-5849

Received: 10 April 2019; Accepted: 15 May 2019; Published: 21 May 2019

Abstract: The working principle of the wave energy converter (WEC) developed at Uppsala University (UU) is based on a heaving point absorber with a linear generator. The generator is placed on the seafloor and is connected via a steel wire to a buoy floating on the surface of the sea. The generator produces optimal power when the translator's oscillations are centered with respect to the stator. However, due to the tides or other changes in sea level, the translator's oscillations may shift towards the upper or lower limit of the generator's stroke length, resulting in a limited stroke and a consequent reduction in power production. A compensator has been designed and developed in order to keep the generator's translator centered, thus compensating for sea level variations. This paper presents experimental tests of the compensator in a lab environment. The wire adjustments are based on online sea level data obtained from the Swedish Meteorological and Hydrological Institute (SMHI). The objective of the study was to evaluate and optimize the control and communication system of the device. As the device will be self-powered with solar and wave energy, the paper also includes estimations of the power consumption and a control strategy to minimize the energy requirements of the whole system. The application of the device in a location with high tides, such as Wave Hub, was analyzed based on offline tidal data. The results show that the compensator can minimize the negative effects of sea level variations on the power production at the WEC. Although the wave energy concept of UU is used in this study, the developed system is also applicable to other WECs for which the line length between seabed and surface needs to be adjusted.

Keywords: wave energy converter; tidal compensation; control system; tides; Wave Hub

1. Introduction

Research on the production of electrical energy from ocean waves has evolved significantly over the years, with the development and testing of various types of wave energy converters (WECs). Different techniques and strategies have been implemented in order to improve the performance of WECs both at the simulation and theoretical level [1–4] and at the sea testing level [5–7]. The analysis and the evaluation of the performance of WEC technologies have also been studied and can be referred to previously-published works; see [8–10]. At Uppsala University, the studied and developed WEC is a heaving-point-absorbing-type converter.

The first version of the sea level compensator from Uppsala University, which was dimensioned for the climate of the Swedish west coast, has been tested [11]. A study by Castellucci [12] analyzed the effect of sea level changes on the absorption of wave energy. The result from the study showed that the implementation of the compensator significantly improved energy production.

This paper focuses on the next-generation tidal compensator for Uppsala University WECs. The tidal compensator is dimensioned to work in the Welsh Sea, which is characterized by a semi-diurnal tidal range of up to 6.6 m [13]. Because of the high tidal range, a higher range tidal compensator for the Uppsala University WEC is needed. In optimal operating conditions, the translator oscillates centered to the mean position of the generator, with the changes in water level (tides) causing the translator to shift towards the top or the bottom of the generator, thereby reducing the efficiency of the WEC.

The compensator consists of a pocket wheel hosting a steel chain and a gearbox. The system is driven by a DC motor rated at 1500 W. A steel chain of 9 m connects the compensator to the buoy line. This chain rolls over the pocket wheel to adjust the length of the buoy line. The gearbox is located between the pocket wheel and a DC motor via a steel shaft. Because of a total load (translator and buoy line) of 10 tonnes, a gearbox with a reduction ratio of 4758 has been used [14]. This can increase the lifting torque (or force) significantly without compromising the ability of the system to adjust the buoy line according to the tidal level.

In order to protect the sensitive parts from corrosion, only the chain and the pocket wheel are allowed to be exposed to the sea water. All three parts (pocket wheel, chain, and shaft) are made of non-corrosive material. The rest of the system is covered by a protective casing. To prevent the sea water from entering the casing, two pairs of U-shaped sealings have been installed.

The real-time experimental test in the lab environment was carried out on the basis of the sea level data at Brofjorden [15]. The sea level at Brofjorden has been chosen as it is the closest measurement station to the Lysekil wave energy research site, located around 12 km to the south. For the adaptation of the system to higher ranges of sea level variation, offline information on tidal levels from a station close to Wave Hub has been analyzed. Offline sea level data from the Newlyn Tidal Observatory obtained from the British Oceanographic Data Centre (BODC) [13] have been used to simulate the compensator's operation at Wave Hub. The Newlyn station is located 22 km south of Wave Hub. The time series analyzed was one year long (from May 2017 until April 2018) with a 15-minute temporal resolution.

This paper presents the simulations and lab experiments on the compensator system based on the sea level information published by the Swedish Meteorological and Hydrological Institute (SMHI) and the BODC. The Methods Section explains the experimental setup of the compensator in the lab environment, followed by the operational test of the compensator. The accuracy of the chain position and the analysis of the power consumption are presented in the Results Section.

2. Methods

This section is divided into three subsections. Section 2.1 describes the communication and control strategy. Section 2.2 provides the experimental setup, describing the sensors and the testing for static and dynamic operation. Section 2.3 models the behavior of the motor to estimate the power consumption.

2.1. Communication and Control Strategy

A server for the compensator system was set up to log the data and to handle the control strategy for the compensator. A simplified block diagram of the communication and control flows is shown in Figure 1.

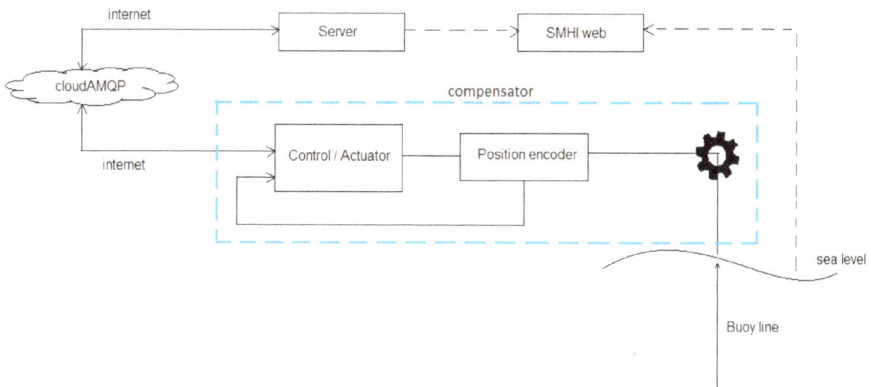

Figure 1. Control flow for the sea level compensation system. The part of the system surrounded by the blue dashed line is shown in Figure 2.

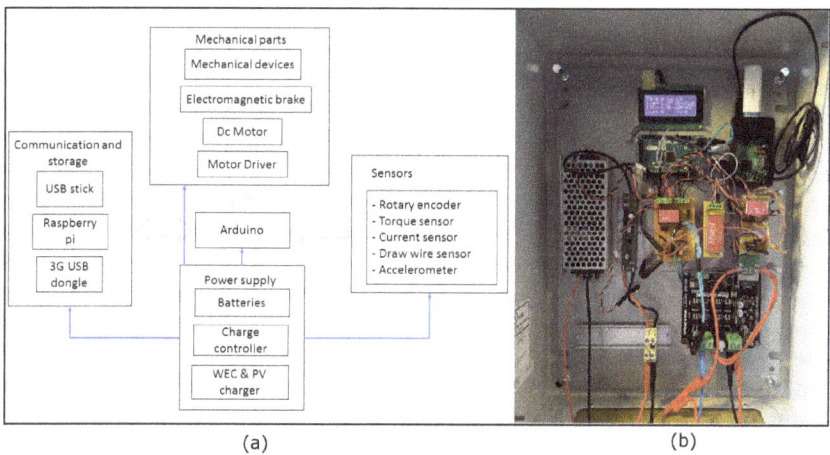

Figure 2. (a) Schematic of the compensator system [16]. (b) Electrical connection installed in the metal cabinet.

The input parameter of the system is the sea level, collected online from the SMHI website [15] (see Figure 3). The observed sea level data published on the website lag between 1 and 2 h. In addition to the observed data, two other types of data series are made available by the SMHI, namely short-forecast data and long-forecast data. The easiest way to control the compensator would be to follow the short-forecast data as the observed data are not published in real time, whereas the forecast data are available ahead of the current time. However, it can be seen from Figure 3 that there is a noticeable variation between the sea level that is observed at the station and the sea level that is forecasted. To minimize this variation, a prediction model using least-squares minimization was employed to predict the current sea level up to 2 h in advance based on the observed sea level data series.

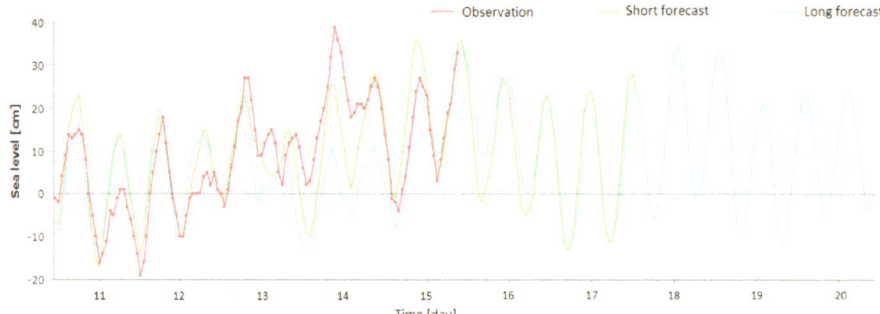

Figure 3. The forecasted and the observed sea level for a period of 10 days (long forecast), 7 days (short forecast), and 5 days (observation) recorded at the Brofjorden measurement station in October 2018.

The prediction of the current sea level data was based on the observational data from the SMHI. Because sea level variations can, in general, be considered to be periodic, a moving frame with a sinusoidal shape for the sea level was used to obtain the data for the next two hours (sea level at the present time). The illustration of this procedure is shown in Figure 4.

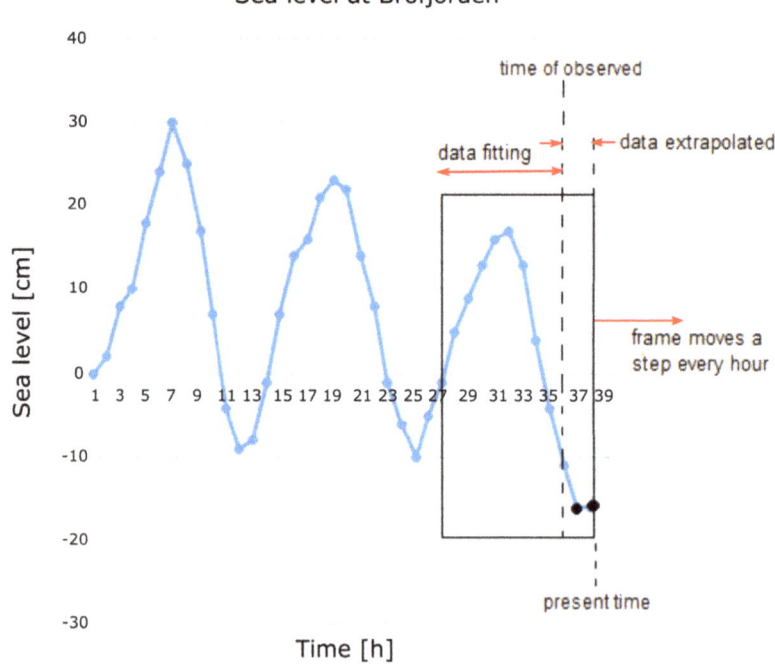

Figure 4. Illustration of 2-h extrapolation based on a sinusoidal fitting for 10 h of observed sea levels.

The rectangular frame shows how 10 observations were used to extrapolate the subsequent data point. The least-squares minimization method [17] was used to fit these 10 observations to get the best fit of a sine function. The acquired sine function was then used to extrapolate the next 2 h of data points every hour (up to the present time) (see Figure 4). The frame was shifted one hour when new observational data were obtained from the SMHI. This process was repeated to predict the sea level data points of the next hour continuously. A simulation to analyze the difference between the extrapolated data and the observed and forecasted data was performed for a period of 19 days.

2.1.1. Server

The system server was placed on land. This was the main component of the communication and control system. The server reads the sea level information from the SMHI website, does the analysis, and makes a prediction of the seal level for up to 2 h in advance based on the SMHI observations. The SMHI updates the observational data on an hourly basis, so the server accesses the website once every hour. The communication between the server and the compensator is done via the cloud service cloudAMQP [18] based on RabbitMQ.

2.1.2. CloudAMQP

CloudAMQP is a cloud server developed by a Swedish tech company. CloudAMQP handles message queues based on the RabbitMQ protocol (open source multi-protocol messaging broker). A basic account from this cloud service with limited connection and queue data is enough to handle the small-sized data exchange between the server and the compensator. For our application, the compensator kept sending the status of the sensor readings (and position information) to the cloudAMQP. These data were then forwarded to the server at Uppsala University. The same protocol was used to send commands regarding the chain's current and future positions from the server to the compensator. The advantage of using this cloud messaging service is its simplicity. The communication between the server and the compensator is asynchronous.

2.2. Experimental Setup

Figure 2 shows the schematic of the components installed in the compensator device. Every component has been explained in detail in [16]. In general, the heart of the control is the Arduino Mega2560 [19], which is linked together with other components to a single-board computer (Raspberry Pi) used for communication via a 3G dongle. The compensator sends the status to the server (via cloudAMQP) at a frequency of 1 Hz. Raspberry-pi can also be accessed manually by VPN using the RealVNC application. However, normally, the automated operation of the compensator cloudAMQP was chosen.

The Arduino Mega2560 was chosen as the main controller due to its low cost and the fact that it is open source and easy to integrate with other systems. As the control system for the compensator does not need high processing power and several input/output ports, the Arduino seems suitable. Raspberry Pi was used as the onboard computer with the purpose of performing data logging and Internet communication. Because of its low power consumption and its potential to control and operate remotely, Arduino has been used for the Internet of Things (IoT) application in real-time data monitoring and management, e.g., the photovoltaic (PV) systems monitoring, as described in [20,21].

2.2.1. Encoder as a Position Sensor

A rotary encoder was used to calculate the rotation of the motor shaft and, subsequently, to calculate the motor rpm and position of the chain. A Hall effect sensor was mounted to face a rotating axis of two permanent magnets with the poles facing opposite directions. Figure 5 shows the setup for the encoder.

The two permanent magnets were installed 90° apart from the rotating coupler in order to obtain a fixed duty cycle of the output signal regardless of the speed of the motor shaft. From this output, the direction and the speed of rotation could be calculated. The sensitivity of the encoder was one pulse per revolution. This was translated to a 0.2-mm resolution of the buoy line position. Figure 6 shows 2 waveform samples from the encoder signal. The output from the encoder was fed to the Arduino for a feedback measurement. Table 1 shows the information obtained from the encoder measurement.

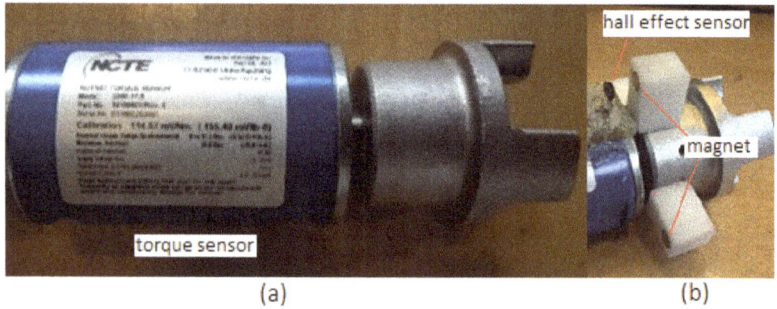

Figure 5. (**a**) Torque sensor with the coupler attached to the shaft. (**b**) Two permanent magnets attached to the coupler, connecting the shaft of the torque sensor to the gearbox. A Hall effect sensor acts as a switching device, which turns on/off when facing the magnets of different poles.

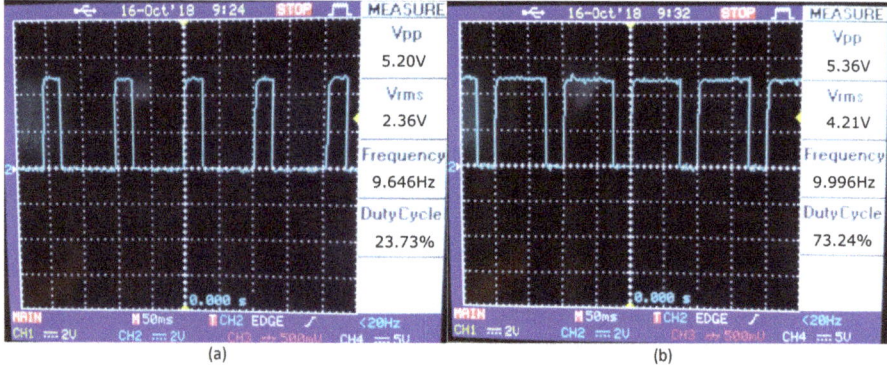

Figure 6. Recorded encoder output. (**a**) Motor clockwise (CW) rotation. (**b**) Motor counter-clockwise (CCW) rotation. The output from the sensor always gives approximately a 25% or 75% duty cycle from the CW or CCW regardless of the speed.

Table 1. Information from the encoder measurements.

Figure 6	Duty Cycle	Direction: Connection Line	Motor rpm	Speed at Chain (mm/s) rpm × 3.5 × 10^{-3}	Linear Position Corresponding to One Motor Revolution
(a)	25%	CW: release	579	2.0	0.2 mm
(b)	75%	CCW: retract	600	2.1	0.2 mm

The position of the magnet was placed so that when the motor turns clockwise (CW), the encoder will always generate approximately a 25% duty cycle. Conversely, when the motor turns counter-clockwise (CCW), the encoder generates signals of a 75% duty cycle. The information of the duty cycle and frequency from the encoder was used to determine the direction, the speed, and the position of the motor when lifting and releasing the load. The position of the chain was saved every time the Raspberry Pi received inputs from Arduino in order to protect the system from data loss, in case it was reset. A secondary position measurement system was also installed: a draw-wire sensor (DWS) was mounted on the low-speed side of the gearbox (12 cm-diameter shaft). The wire was wound over the shaft and was able to give the absolute position value when needed. However, the draw wire sensor was not as accurate as the encoder and can be used as a backup if the encoder system fails.

2.2.2. Force Measurement

The compensator was designed to be able to bear the weight of the translator and the connection line, which resulted in a total of 10 tonnes. A torque sensor (Figure 5) [22] was installed to measure the torque and estimate the force exerted on the chain. The nominal range for this torque sensor was 17 Nm, which translates to about 83 kNm at the output of the gearbox. The torque sensor was installed at the coupling between the DC motor's shaft and the gearbox.

In order to test the torque sensor, experiments were carried out. The lab environment with the setup is shown in Figure 7. Doing dynamic tests on the compensator using the real load conditions is complicated and poses safety risks; hence, static tests were performed using a load cell mounted on the floor. The load cell shown in Figure 7 was rated at 25 tonnes.

For the static tests, one side of the load cell was connected to the chain and the other end to the floor, as shown in Figure 7. During the test, the load cell was pulled by the chain to measure the pulling force on the compensator. The result of this experiment, shown in Figure 8, was used to calibrate the torque sensor: the force measured by the load cell should be equal to the force calculated from the output of the torque sensor. Even though the gearbox turn ratio was high, the pull of the load on the torque sensor cannot be ignored. From the test, the instantaneous force applied by the chain was represented by the pull of the spring in the releasing-load direction (in the dynamic test) when the motor was stopped, and from the results, it was shown that the comparison of the forces measured at the spring and forces obtained from the torque sensor matched to an acceptable accuracy [16].

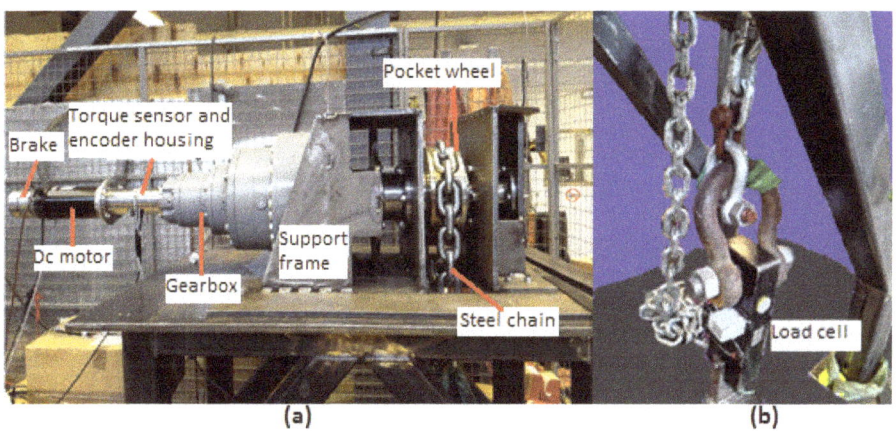

Figure 7. (a) The compensator system with the parts labeled in the image. (b) The steel chain pulls the load cell.

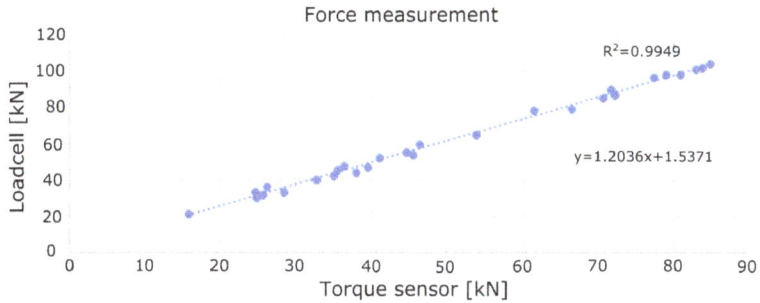

Figure 8. Result from the torque sensor calibration test.

2.3. Power Consumption Estimation

The power consumption is one of the constraints for the system's operation. The performance analysis of the system is focused on estimating the power consumption. To do so, the motor behavior was modeled in steady-state operation. Because the rate of sea level change is low, a fast response of the system is not necessary. The total power consumption was due to (a) the power consumption of the motor and the brake release when moving the chain and (b) the standby power necessary for monitoring and communication. The brake was powered only when the motor moved. At 24 V, the brake draws around 2 A current. Thus, the energy required for brake release is:

$$E_b = 48t, \qquad (1)$$

where t is the time it takes to complete the operation. The response of the motor was analyzed only in the steady-state condition. An experiment to characterize the system was performed by lifting a 35-kN load. The details of the test can be found in [16].

The relationship between the current, I, and the torque, T, can be obtained by measuring the torque when different currents are fed into the motor. This dynamic test to find the relation of I and T is plotted and shown in Figure 9. From the figure, it is noticeable that, in general, the torque was directly proportional to the current fed in. Ideally, the linear fit shown in Figure 9 should cross the origin, but in the experiment, a small current was required to overcome the internal friction of the motor before it started to turn. A static test was easier to determine the torque constant, but this dynamic test was more accurate to estimate the relation between I and T accounting for the effect of dynamic loses. The linear fit in Figure 9 shows the relation of I and T as:

$$T = 0.2045I - 0.1428. \qquad (2)$$

With the same test, by considering the main loses of the DC motor to be the copper loses, where the resistance $R_a = 0.32$ Ohm, the relationship between the input power to the motor and the output power calculated with the torque measurements can be obtained. The results are shown in Figure 10.

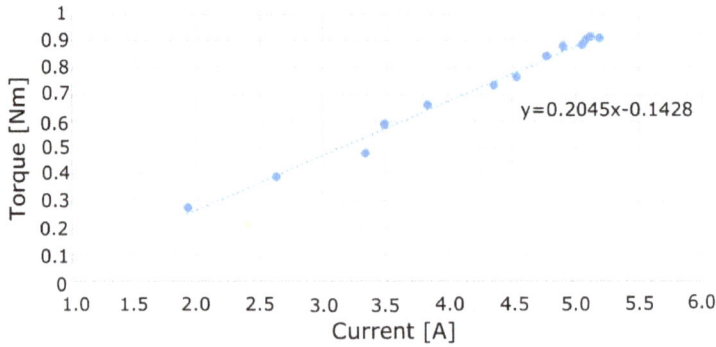

Figure 9. The torque measured at the output shaft of the DC motor with respect to the current fed in.

The significant losses of the system were considered to be copper losses, which were proportional to the square value of the current in the rotor windings. However, from the experiment, the efficiency of the motor, η_m, after the copper losses had been accounted for, was found to be around 95 percent. The output power of the motor, $T\omega$, measured at the output motor shaft can be estimated as:

$$P_{out} = P_{in}\eta_m. \qquad (3)$$

Expanding Equation (3) yields:

$$T\omega = (VI - I^2 R_a)\eta_m. \tag{4}$$

From Equation (4), $T\omega$ can also be presented as the power exerted on the motor shaft from the perspective of the load. At steady state,

$$T\omega \eta_s = P_{mech} + P_{load}, \tag{5}$$

where P_{mech} is the power needed to operate the system without load. P_{load} is the power needed to lift the translator, and η_s is the dynamic efficiency of the compensator. The value of P_{mech} has been obtained from experimental (no load) tests. In order to estimate the power when the translator is lifted (or released) at a constant speed, ω_a, Equation (5) can be presented as:

$$(VI - I^2 R_a)\eta = T\omega_a|_{T\,at\,\omega_a} + F_{load} v|_{v\,at\,\omega_a}, \tag{6}$$

where $\eta = \eta_s \eta_m$ is the efficiency of the system, F_{load} is the weight of the translator (100 kN), and v is the linear speed of the chain at ω_a. From Equation (6), the total power on the right-hand side can be estimated and is equal to $T_{total}\omega_a$. Because ω_a is known and the torque, T_{total}, of the motor is proportional to the current I, the value of the required current can be obtained; the required voltage can subsequently be determined by equating $T_{total}\omega_a$ to the left-hand side of Equation (6). Then, the power consumption to run the motor with load F_{load} at speed ω_a can be estimated.

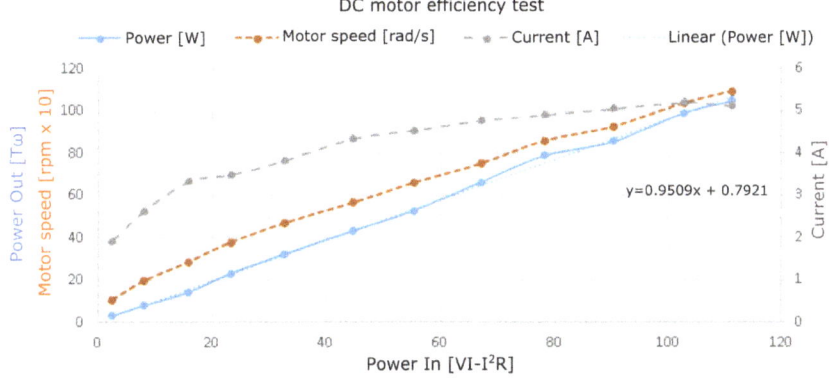

Figure 10. The relationship between the power fed in and the power put out, as measured from the DC motor, to determine the other losses after the copper losses ($I^2 R a$) have been accounted for.

3. Results

Section 3.1 will present the simulation and experimental results of the compensator positioning for Brofjorden's sea level. The effects on power consumption from different control strategies is presented in Section 3.2. Finally, the selected control strategy is adopted for Wave Hub and its effect on power consumption is analyzed in Section 3.3.

3.1. Compensator Positioning

Figure 11 shows the result from the simulation of the compensator position together with the sea level information from the SMHI for a duration of 19 days.

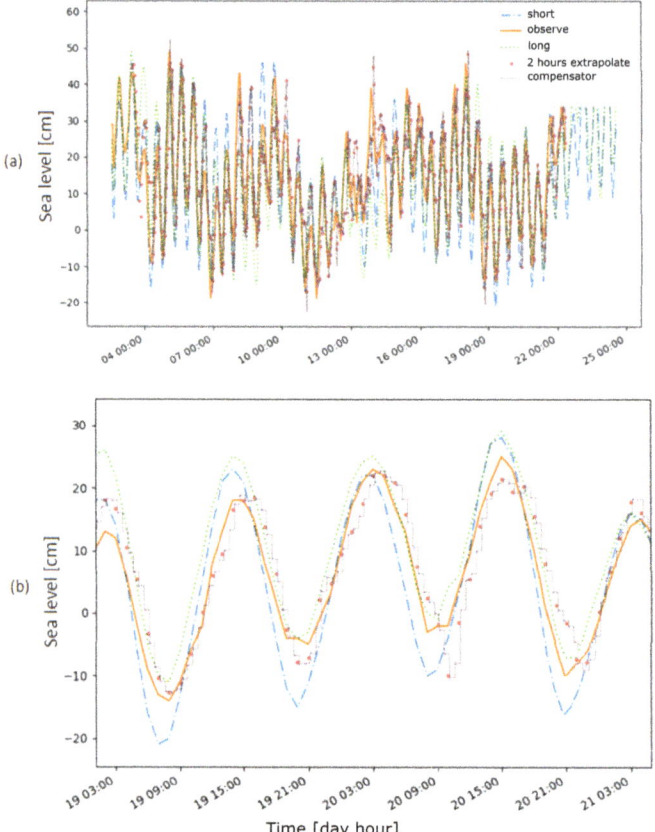

Figure 11. (a) Two-hour prediction from the observational data compared to long- and short-forecast data in October 2018 (available from SMHI). (b) A zoom in of two days of data.

Table 2 shows the error analysis of long-forecast, short-forecast, and extrapolated sea levels compared to the observed data from the SMHI. The result of the root mean squared error (RMSE) analysis shows that the RMSE of the extrapolation from the observed data gave the smallest error, as compared with the error calculated for the short- and long-forecast data. Therefore, the first method was used to calculate the new position of the chain. Although the extrapolation left a margin of error, this will not significantly influence the power production, as discussed in a previous study on the impact of tidal levels on the WEC power production [12].

Table 2. Root mean squared error (RMSE).

Data Type	2-h Extrapolation	Short Forecast	Long Forecast
RMSE (cm)	7.4	8.1	9.9

Time analysis: 3 October 2018, 1100–22 October 2018, 0900 (19 days).

The user can set the parameters, as shown in Figure 12. By setting the interval time, the server will run fully automatically, reading SMHI information, continuously extrapolating the new position every hour, and sending commands to the compensator. There are several other parameters available on the GUI for manual positioning and safety.

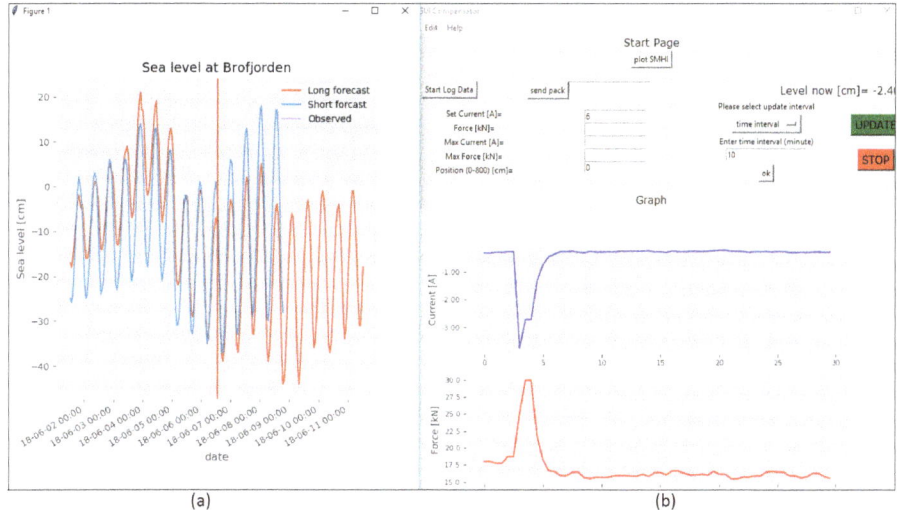

Figure 12. (a) Sea level at Brofjorden. (b) GUI for the control settings.

Figure 13 shows the experimental results from a chain positioning test. The experiment was held in the lab environment without the translator load for a duration of two days. The purpose of the experiment was to evaluate the communication and the positioning control of the system. The result shows that the compensator could follow the sea level changes in real-time operation.

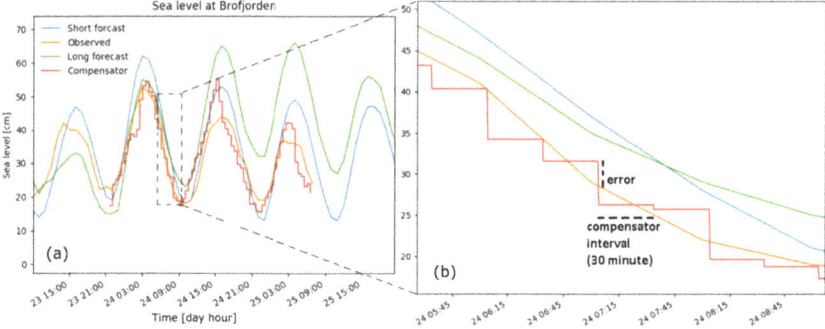

Figure 13. (a) Experimental results from a positioning test of the chain in October 2018. (b) Zoom out of the area marked by the dotted rectangle in (a).

3.2. Power Consumption with Different Control Strategies

As the compensator will operate as a self-powered device [23], one of the constraints in the operation is to minimize the power consumption while achieving the targets for positioning at minimal error. The plot in Figure 14 shows the simulation results of running the compensator based on the data series of Figure 11. The result shows the estimated energy required to adjust the chain position. The load calculated as the translator and the connection line weights was estimated to be 100 kN. In general, the faster the compensator system is allowed to turn, the less energy is required for the operation and the higher voltage is needed to power the motor. However, the device was chosen to be powered by two 12-V, 100-Ah batteries connected in series. Figure 15 shows the estimated voltage required to turn the motor at the designated rpm (left, vertical axis). From this figure, it can be concluded that the motor should be operated at the highest battery capacity of 24 V to turn the motor at the highest speed (800 rpm) and with high efficiency. With reference to Figure 14, it is noticeable that at 800 rpm, the energy required to operate the system is already at a minimum.

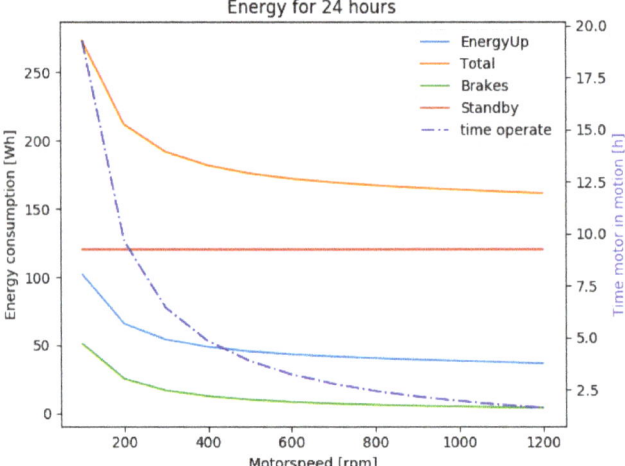

Figure 14. The analysis of the energy required by the motor and the brake and the standby power for the 19-day period analyzed (as in Figure 11). The dashed blue line shows the total time required to turn the motor. The green line shows the energy required to release the brake when the motor turns. The red line shows the standby energy, which is approximately the same throughout the test period.

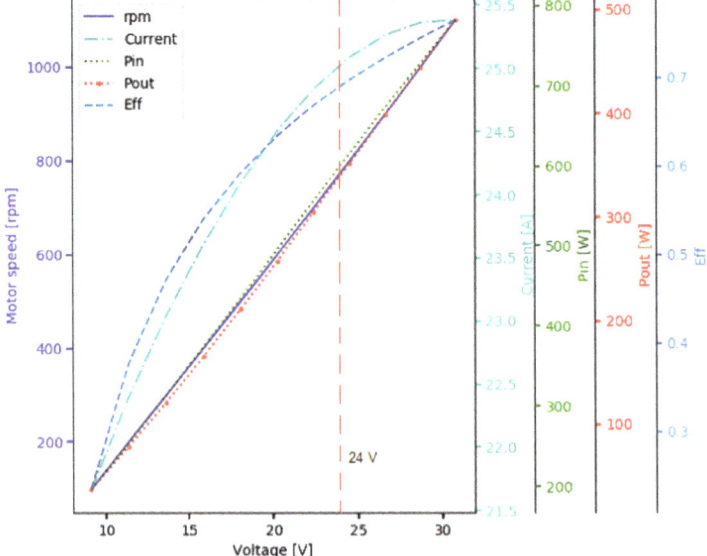

Figure 15. The motor behavior with respect to speed and its corresponding parameters when the motor is powered with different voltage levels and the applied load is set to 100 kN. The red dashed line shows the level of the nominal voltage supplied by the batteries.

3.3. Adaptation to Wave Hub

The analysis of the estimated power consumption of the compensation system at Wave Hub is shown in Figure 16. The simulation compared the control strategy for four different time intervals between each compensating operation. 15-, 30-, 60-, and 120-min intervals were chosen for the study. In each case, the compensator will start to compensate until it reaches the calculated position (extrapolated) of the sea level. Table 3 tabulates the RMSE values as the performance evaluation of the compensator position; see the dashed line (labeled with "error" and "compensator interval") in Figure 13b. The power consumptions and the RMSEs were analyzed for a period of one year (2017–2018). The result shows that the power consumption of the system when it runs at its full capacity of 24 V, which corresponds to 800 rpm (see Figure 15), was around 600–640 Wh. As the batteries were rated 100 Ah, 24 V, the system should last for three days without charging.

Table 3. RMSE between compensator position and sea level for different control strategies.

Interval Time (minutes)	RMSE (cm)
15	17
30	34
60	67
120	130

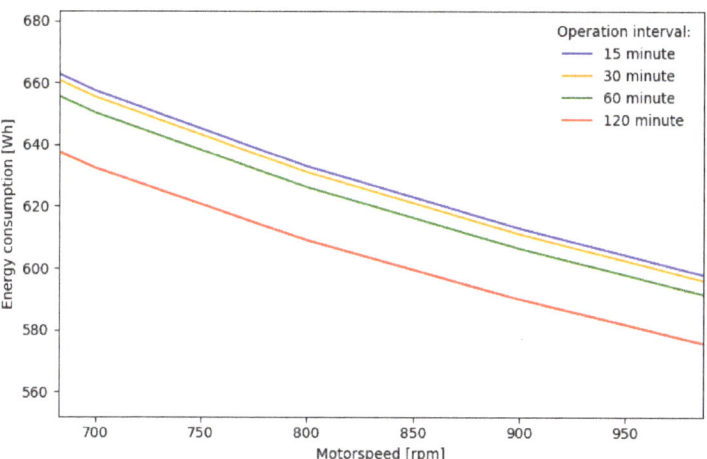

Figure 16. Estimated energy consumption for 24 h with different compensating time intervals for the operation at Wave Hub.

4. Discussion

Due to the high turn ratio of the gearbox (1:4758), one pulse per revolution of the motor is enough to get high accuracy on the chain position. As the number of the revolution of the shaft is high, the encoder suitable for this application is the incremental type, with the ability to restore the last position (saved in Raspberry Pi) after a reset. Moreover, the custom protocol as it is currently designed has facilitated the system well: the slow encoder output frequency, yet high accuracy (due to the high turn ratio), and the mechanism of detecting turn direction (CCW/CW) on every encoder's pulse.

The highest pulling force tested in the lab was 104 kN. The result plotted in Figure 8 shows that the force measurements from the torque sensor were reliable. Adjustments were made to calibrate and remove the sensor offset. In order to test the system in real time, we removed the load (springs) for safety reasons. Moreover, the main goal for the positioning test was to evaluate the control strategy and communication for a long duration, from the collection of sea level data to moving the chain to the new, desired position. Due to the slow change in sea level at the measurement station of Brofjorden, the lag between data collection and the adjustment of the connection line can be neglected.

The approach of utilizing a series of 10 h of sea level information to predict the next two hours was based on the understanding that changes in sea level follow a semi-diurnal trend with an M2 constituent period of about 12 h and 25 min. Based on this, the previous 10 h of data were used to extrapolate the data points of the next two hours and to get rid of the effects due to low-frequency variations. This method was also used to simplify the calculations. The analysis of the RMSE values calculated over 19 days of data led to an error between predictions and observations of 7.4 cm (see Table 2). This RMSE value was found to be lower than the RMSE of short- and long-forecast data; hence, it was selected for calculating the new desired position of the chain.

Attempts to lift a heavy load dynamically have been done in [16], but only loads of up to 35 kN could be lifted for safety reasons in the lab environment. However, there were many uncertainties due to the configuration of the setup and the accuracy of the measurements. Our test improved the understanding of the system and the DC motor behavior based on the input signal. It also contributed to dimensioning the power supply system and evaluating power consumption. From the plot in Figure 14, it can be seen that the faster the motor, the less total energy is required for the operation. However, this will be smoothed out because of the reduced efficiency of the motor at high speed (and current). Looking at the behavior of the curve of total energy consumption, we decided that the motor should run at a steady state of 800 rpm. This would correspond to a standard battery supply of 24 V. The relationship between the speed of the motor and the voltage supplied is shown in Figure 15.

The result of the analysis of the speed and operating intervals for Wave Hub shows that the effect of the interval control on the power consumption was minimal. All intervals tested consumed roughly the same amounts of energy. One thing to consider in this respect is to put a limit on the interval (on/off time) so that the error of the compensator can be kept at a minimum to optimize the power production of the WEC. With reference to [12,24], the negative effects of the changes in tidal level on the power production were in general minimal when the variation in the tides were in the range of 0.2 m at Wave Hub. As can be inferred from Table 3, a time interval of 15 min (or shorter) was suitable to keep the error of the compensator and the sea level the whole time smaller than 0.2 m. However, the effect of the sea level change on the power production will also depend on the stroke length of the generator and the wave climate. At smaller wave heights, the negative effect of sea level change is minimal, and thus, the time interval for the compensator can be longer while maintaining the power production.

The input (sea level) to the system is currently obtained from the observatory-station's sea level data available online from SMHI. This strategy can be robust considering it suits different types of WEC devices regardless of whether the WEC (or a wave farm) has its local sea level measurement. However, if a WEC has its position measurement installed, e.g., translator position in the generator, this position input can be used for a more accurate compensation (directly compensates for translator's shift) compared to the measured (or forecast) data from observatory station, which is usually situated away from the wave farm.

5. Conclusions

A remotely-controlled sea level compensator for WECs was designed, built, and evaluated. The experiment was held in the lab with the input to the system being the observed sea levels at Brofjorden observatory station, collected online from SMHI. Additionally, offline sea level data were analyzed, and the device operations were simulated to evaluate the control strategy and the compensator behavior for the operation of the device in a higher tidal range, e.g., Wave Hub.

The experimental tests in the lab have shown that the integration of the control and the communication systems worked reasonably well. As a self-powered device, the system had the right dimensions to be capable of lifting the load with minimal energy, as demonstrated by the analysis of the power consumption. The daily power consumption was not affected by the time interval selected for the operation, at least for the time period simulated. However, the power consumption was more influenced by the permitted error variations (RMSE) of the compensator for optimal power production on the side of the main WEC: the larger the RMSE for the compensator positioning resulted in lower power production at the main WEC. For Wave Hub to maximize the power production, the time interval between the adjustments should not be longer than 15 min at a motor speed of approximately 800 rpm.

Author Contributions: M.N.A. contributed to the development of the device, designing the experimental setup and control system, analyzing the result, and writing the article. V.C. contributed to the experimental setup and analyzing the result. J.A. contributing technical support. R.W. supervised the work and provided financial support. All co-authors participated in writing the article.

Funding: This research was funded by Swedish Energy Agency grant number 2016-002062.

Acknowledgments: M.N.A. is funded by the Ministry of Education of Malaysia. J.J. Perez-Loya and D. Salar are acknowledged for their help with the experimental setup. Thanks go to the Swedish Energy Agency and Ångpanneföreningen for their financial support.

Conflicts of Interest: The authors declare no conflict of interest.

Abbreviations

The following abbreviations are used in this manuscript:

AMQP	Advanced Message Queuing Protocol
BODC	British Oceanographic Data Centre
CCW	Counter-clockwise
CW	Clockwise
DC	Direct current
DWR	Draw wire sensor
GUI	Graphical user interface
IoT	Internet of Things
M2	Principle lunar semi-diurnal tide
MQ	Messaging queue
PV	Photovoltaic
RMSE	Root mean square error
SMHI	Swedish Meteorological and Hydrological Institute
UU	Uppsala University
VPN	Virtual private network
WEC	Wave energy converter

References

1. Engström, J. Hydrodynamic Modelling for a Point Absorbing Wave Energy Converter. Ph.D. Thesis, Electricity, Department of Engineering Sciences, Technology, Disciplinary Domain of Science and Technology, Uppsala University, Uppsala, Sweden, 2011.
2. Yavuz, H.; McCabe, A.; Aggidis, G.; Widden, M.B. Calculation of the performance of resonant wave energy converters in real seas. *Proc. Inst. Mech. Eng. Part M* **2006**, *220*, 117–128. doi:10.1243/14750902JEME44. [CrossRef]
3. Eriksson, M.; Isberg, J.; Leijon, M. Hydrodynamic modelling of a direct drive wave energy converter. *Int. J. Eng. Sci.* **2005**, *43*, 1377–1387. doi:10.1016/j.ijengsci.2005.05.014. [CrossRef]
4. Bozzi, S.; Miquel, A.; Antonini, A.; Passoni, G.; Archetti, R. Modeling of a Point Absorber for Energy Conversion in Italian Seas. *Energies* **2013**, *6*, 3033–3051, doi:10.3390/en6063033. [CrossRef]
5. Lejerskog, E.; Boström, C.; Hai, L.; Waters, R.; Leijon, M. Experimental results on power absorption from a wave energy converter at the Lysekil wave energy research site. *Renew. Energy* **2015**, *77*, 9–14, doi:10.1016/j.renene.2014.11.050. [CrossRef]
6. Liang, C.; Ai, J.; Zuo, L. Design, fabrication, simulation and testing of an ocean wave energy converter with mechanical motion rectifier. *Ocean. Eng.* **2017**, *136*, 190–200, doi:10.1016/j.oceaneng.2017.03.024. [CrossRef]

7. Waters, R.; Stålberg, M.; Danielsson, O.; Svensson, O.; Gustafsson, S.; Strömstedt, E.; Eriksson, M.; Sundberg, J.; Leijon, M. Experimental results from sea trials of an offshore wave energy system. *Appl. Phys. Lett.* **2007**, *90*, 034105, doi:10.1063/1.2432168. [CrossRef]
8. Rusu, E. Evaluation of the Wave Energy Conversion Efficiency in Various Coastal Environments. *Energies* **2014**, *7*, 4002–4018, doi:10.3390/en7064002. [CrossRef]
9. Belibassakis, K.; Bonovas, M.; Rusu, E. A Novel Method for Estimating Wave Energy Converter Performance in Variable Bathymetry Regions and Applications. *Energies* **2018**, *11*, 2092, doi:10.3390/en11082092. [CrossRef]
10. Silva, D.; Rusu, E.; Soares, C. Evaluation of Various Technologies for Wave Energy Conversion in the Portuguese Nearshore. *Energies* **2013**, *6*, 1344–1364, doi:10.3390/en6031344. [CrossRef]
11. Castellucci, V.; Abrahamsson, J.; Kamf, T.; Waters, R. Nearshore Tests of the Tidal Compensation System for Point-Absorbing Wave Energy Converters. *Energies* **2015**, *8*, 3272–3291, doi:10.3390/en8043272. [CrossRef]
12. Castellucci, V.; Eriksson, M.; Waters, R. Impact of Tidal Level Variations on Wave Energy Absorption at Wave Hub. *Energies* **2016**, *9*, 843, doi:10.3390/en9100843. [CrossRef]
13. British Oceanographic Data Centre. Available online: https://www.bodc.ac.uk (accessed on 11 May 2016).
14. Bonfiglioli. Available online: http://www.bonfiglioli.com (accessed on 21 February 2017).
15. Swedish Meteorological and Hydrological Institute. Available online: https://www.smhi.se (accessed on 8 March 2018).
16. Ayob, M.N.; Castellucci, V.; Abrahamsson, J.; Svensson, O.; Waters, R. Control strategy for a tidal compensation system for wave energy converter device. In Proceedings of the 28th International Offshore and Polar Engineering Conference, Sapporo, Japan, 10–15 June 2018; pp. 808–812.
17. scipy.org. Available online: https://docs.scipy.org/doc/scipy/reference/generated/scipy.optimize.curve_fit.html (accessed on 9 April 2019).
18. CloudAMQP. Available online: https://www.cloudamqp.com (accessed on 9 April 2019).
19. ARDUINO MEGA 2560. Available online: https://store.arduino.cc/arduino-mega-2560-rev3 (accessed on 24 January 2018).
20. Lopez-Vargas, A.; Fuentes, M.; Vivar, M. IoT Application for Real-Time Monitoring of Solar Home Systems Based on Arduino™ With 3G Connectivity. *IEEE Sensors J.* **2019**, *19*, 679–691, doi:10.1109/JSEN.2018.2876635. [CrossRef]
21. Paredes-Parra, J.M.; García-Sánchez, A.J.; Mateo-Aroca, A.; Molina-Garcia, A. An Alternative Internet-of-Things Solution Based on LoRa for PV Power Plants: Data Monitoring and Management. *Energies* **2019**, *12*, 881, doi:10.3390/en12050881. [CrossRef]
22. NCTE Torque sensor. Available online: https://www.elfa.se/Web/Downloads/_t/ds/serie2000_eng_tds.pdf (accessed on 9 April 2019).
23. Ayob, M.; Castellucci, V.; Göteman, M.; Widén, J.; Abrahamsson, J.; Engström, J.; Waters, R. Small-Scale Renewable Energy Converters for Battery Charging. *J. Mar. Sci. Eng.* **2018**, *6*, 26, doi:10.3390/jmse6010026. [CrossRef]
24. Tyrberg, S.; Waters, R.; Leijon, M. Wave Power Absorption as a Function of Water Level and Wave Height: Theory and Experiment. *IEEE J. Ocean. Eng.* **2010**, *35*, 558–564, doi:10.1109/JOE.2010.2052692. [CrossRef]

 © 2019 by the authors. Licensee MDPI, Basel, Switzerland. This article is an open access article distributed under the terms and conditions of the Creative Commons Attribution (CC BY) license (http://creativecommons.org/licenses/by/4.0/).

Article

Fault Tolerant Control of DFIG-Based Wind Energy Conversion System Using Augmented Observer

Xu Wang and Yanxia Shen *

Key Laboratory of Advanced Process Control for Light Industry, Jiangnan University, Wuxi 214122, China; 6161920009@vip.jiangnan.edu.cn
* Correspondence: shenyx@jiangnan.edu.cn; Tel.: +86-138-8186-7517

Received: 23 December 2018; Accepted: 31 January 2019; Published: 13 February 2019

Abstract: An augmented sliding mode observer is proposed to solve the actuator fault of an uncertain wind energy conversion system (WECS), which can estimate the system state and reconstruct the actuator faults. Firstly, the mathematical model of the WECS is established, and the non-linear term in the state equation is separated as the uncertain part of the system. Then, the states of the system are augmented, and the actuator fault is considered as part of the augmented state. The augmented sliding mode observer is designed to estimate the system state and actuator fault. A robust fault-tolerant controller is designed to ensure the reliable input of the WECS, maintain the stability of the fault system and maximize the acquisition of wind energy. The numerical simulation results verify the effectiveness of the control strategy.

Keywords: wind energy conversion system; augmented sliding mode observer; fault reconstruction; fault-tolerant control

1. Introduction

Wind power generation is the most mature and promising form of new energy generation [1]. The wind energy conversion system (WECS) is an important part of the wind power generation system. It is generally located in the complex terrain and harsh climate environment such as mountain islands. The WECS is prone to frequent faults, seriously affecting the performance of the wind power system, and even causing paralysis of the system, resulting in incalculable losses. Therefore, it is of great practical significance to improve the reliability and security of the WECS [2,3].

There are many control methods currently applied to fault-tolerant control of the WECS. Observer-based fault diagnosis and fault-tolerant control have received extensive attention [4]. Reference [5] estimates the fault of fan pitch actuator by combining a disturbance compensation device with a controller, then modifies the pitch control law appropriately to achieve fault tolerant control comparable to that without fault. An adaptive active fault-tolerant fuzzy controller was designed by using multi-observer switching control strategy to ensure the stability of the WECS, considering the interaction of parameter uncertainties and sensor faults in [6]. Reference [7] proposed a fuzzy reference adaptive control, which adapts the parameters to achieve fault-tolerant control in the case of uncertain system potential faults, so as to adjust the generator torque value. In [8], the convex decomposition theory is used to transform the non-linear model of the fan into the linear model, and the state feedback method is used to obtain the fault-tolerant control of the virtual actuator. In [9], an adaptive fault observer is constructed to diagnose the transmission faults of the WECS and to implement fault-tolerant control. Due to the strict design conditions of traditional observers, the scope of application is limited. High-order sliding mode control strategy is also widely used [10–13]. The high-order sliding mode based on DFIG (doubly fed induction generator) used in [10] as an improved scheme to deal with the classical sliding mode chattering problem, which is robust to external disturbances. Reference [11] adopted maximum power point tracking, optimal fault adaptive tracking and adaptive robust non-linear control combined with high-order sliding mode to control open-circuit fault of generator. Reference [12] presented a second-order sliding mode control based on DFIG wind power generation system, and controls the wind power generation system according to the reference value given by Maximum Power Point Tracking, so as to obtain the maximum power extraction.

A new state variable is composed of input and state variables to form a singular system, which provides an idea for the method of unknown input observer for nonlinear systems [14]. In [15], the output noise and the state variables of the original system are combined into a new generalized system, and a generalized sliding mode observer is designed for the system. Then, H_∞ is used to guarantee the robustness and estimate the output noise. An extended sliding mode observer is designed to estimate external disturbances and system states simultaneously, which widens the application scope of fault diagnosis observer in [16]. Reference [17] reduced the influence of process disturbance by constructing augmented state vector composed of system states and related faults, and estimating system states and related faults. In [18], the discrete linear model is used to design the sliding mode controller. The stability and robustness of the nonlinear system are improved by adding discrete operators to improve the discrete sliding mode controller. A new design method of augmented fault diagnosis observer is proposed in [19], which separates the observer from the output feedback fault-tolerant device and simplifies the design process. In [20], the augmented system, unknown input fuzzy observer and linear matrix inequality are combined to design robust fault estimation and fault tolerance control approach for T-S fuzzy systems, which are applied to 4.8-MW wind turbines system.

In practice, the phenomena of abrupt disturbance, sensor faults and actuator faults are very common, and further studies are urgently needed. From the above research, it can be seen that for wind power generation system, fault reconfiguration and fault tolerance of design robust sliding mode observer can be achieved by enlarging the system, which reduces the knowledge and experience requirements of the system. The design process is simple and easy to implement, and improves the robustness of the system.

In this paper, an augmented sliding mode observer is proposed to solve the actuator fault of uncertain WECS. By dividing the non-linear term into a constant matrix and an uncertainty matrix, and augmenting the system state, the actuator fault is augmented as a part of the system state, and an augmented sliding mode observer is constructed. The equivalent output control method is used to reconstruct the fault without affecting the state estimation. The active fault-tolerant controller is designed to ensure the reliable input of the WECS. Finally, the proposed method is validated on the wind turbine model.

2. Mathematical Modeling of Double-fed WECS

The WECS is mainly composed of wind turbine, transmission system, generator, AC-DC converter, power grid and so on. Wind turbines convert the captured wind energy into mechanical energy, drive the doubly-fed induction motor to rotate through the transmission link, and transmit the generated energy to the power grid [21]. The DFIG-based WECS is shown in Figure 1.

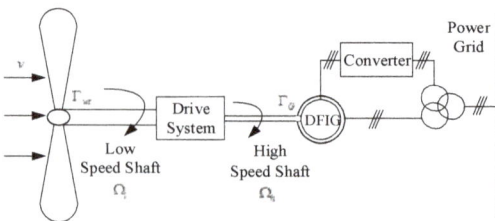

Figure 1. Wind energy conversion system (WECS) based on Doubly fed induction generator (DFIG).

According to Betz' Law, assuming that the wind turbine is in an ideal state, the mechanical power obtained by the wind turbine is as follows [22]:

$$P_{wt} = 0.5\pi R^2 \rho v^3 C_p(\lambda, \beta) \qquad (1)$$

where, P_{wt} is the mechanical power captured by the wind turbine, ρ is the air density, v is the wind speed, R is the fan blade length; $C_p(\lambda,\beta)$ is the wind energy conversion coefficient, which is a function of tip velocity ratio λ and pitch angle β. λ is the ratio of tip speed to wind speed, that is $\lambda = \Omega_l(R/v)$, where Ω_l is the angular velocity of the wind turbine rotor, that is, the low speed axis.

Equation (1) shows that when the wind speed is constant, the mechanical power captured by the wind turbine is only related to $C_p(\lambda,\beta)$. If the pitch angle β of the wind turbine remains unchanged, the wind energy conversion coefficient C_p is only related to the tip speed ratio λ. For different types of wind turbines, there is an unique optimal tip speed ratio λ to ensure the best wind energy conversion coefficient C_p and achieve maximum wind energy capture.

The wind torque generated by the wind wheel is as follows:

$$\Gamma_{wt} = \frac{P_{wt}}{\Omega_l} = 0.5\pi \rho v^2 R^3 C_\Gamma(\lambda, \beta) \qquad (2)$$

where, $C_\Gamma(\lambda,\beta) = C_p(\lambda,\beta)/\lambda$ is the torque coefficient.

The transmission system of the WECS is mainly composed of the wind turbine rotor, low speed shaft, variable speed gear, high speed shaft and generator rotor. The mechanical energy from the fan drives the low-speed shaft to rotate, and electromagnetic torque is produced. Through the gear box transformation, the lower speed of the blade is increased to a higher speed, which is transmitted to the generator rotor to drive the DFIG to rotate, and the electric energy is output to the power grid. For simplicity, rigid models are generally used for the connection between high-speed and low-speed axles. The dynamic equations are as follows:

$$J_h \frac{d\Omega_h}{d_t} = \frac{\eta}{i_0}\Gamma_{wt} - \Gamma_G \qquad (3)$$

$$J_t \frac{d\Omega_l}{d_t} = \Gamma_{wt} - \frac{i_0}{\eta}\Gamma_G \qquad (4)$$

where, Ω_h is the rotor speed (high speed shaft) of the generator, that is $\Omega_l = i_o \times \Omega$, i_o is the gear transmission speed ratio. Γ_G is the electromagnetic torque of the generator, η is the transmission efficiency, J_h is the inertia of the high-speed axis, J_t is the inertia of the low-speed axis.

Based on the power coefficient and the optimal tip speed ratio λ, considering that the generator is in an ideal state, the state equation of the WECS [23] is modeled as follows:

$$\begin{cases} \dot{\Omega}_h(t) = \frac{\Gamma_{wt}(i_0 \cdot \Omega_h, v)}{i_0 \cdot J_t} - \frac{\Gamma_G}{J_t} \\ \dot{\Gamma}_G(t) = -\frac{\Gamma_G}{T_g} + \frac{\Gamma^*_{ref}}{T_g} \end{cases} \qquad (5)$$

where, Γ^*_{ref} is the reference value of the electromagnetic torque of the generator and T_G is the electromagnetic time constant. Taking Ω_h and Γ_G as state vectors, the state equation of the WECS is obtained as shown in (6):

$$\begin{cases} \dot{x}(t) = A'x(t) + Bu(t) \\ y(t) = Cx(t) \end{cases} \qquad (6)$$

where, $\dot{x}(t) = [\Omega_h, \Gamma_G]^T$, $u(t) = \Gamma^*_{ref}$, $A' = \begin{bmatrix} \frac{\Gamma_{wt}(i_0\Omega_h,v)}{i_o J_t \Omega_h} & -\frac{1}{J_t} \\ 0 & -\frac{1}{T_G} \end{bmatrix}$, $B = \begin{bmatrix} 0 \\ \frac{1}{T_G} \end{bmatrix}$, $C = \begin{bmatrix} 1 & 0 \\ 0 & 1 \end{bmatrix}$, $x(t)$ is the state vector, $u(t)$ is the input vector, $y(t)$ is the output vector.

3. Actuator Fault Model of the WECS

Actuator faults are generally caused by wear and tear of gears in gearboxes, wear and deformation of bearing tooth surfaces, and are affected by some uncertainties of the system. Considering these faults of the WECS [24], the system can be described as:

$$\begin{cases} \dot{x}(t) = A'x(t) + Bu(t) + Df_a(t) \\ y(t) = Cx(t) \end{cases} \quad (7)$$

where, $x \in R^n$ is the state variable, $u \in R^m$ is the input vector and $y \in R^p$ is the measurable output vector, $f_a \in R^q$ represents an unknown but bounded actuator fault of the system, $A' \in R^{m \times n}$, $B \in R^{n \times m}$, $C \in R^{p \times n}$, $D \in R^{n \times q}$. The system matrix A' is split into the form of the sum of an uncertain matrix and a constant matrix, that is $A' = \Delta A$, $A = \begin{bmatrix} 0 & -\frac{1}{J_t} \\ 0 & -\frac{1}{T_G} \end{bmatrix}$, $\Delta A = \begin{bmatrix} \frac{\Gamma_{wt}(i_0\Omega_h,v)}{i_0 J_{ty}\Omega_h} & 0 \\ 0 & 0 \end{bmatrix}$ where $\Delta Ax = Md(x,u,t) = \begin{bmatrix} \frac{\Gamma_{wt}(i_0\Omega_h,v)}{i_0 J_t} & 0 \end{bmatrix}^T$, where $d(x,u,t) \in R^h$, $M \in R^{n \times h}$. $d(x,u,t)$ is regarded as an unknown input disturbance of the system, and Equation (7) can be converted into (8):

$$\begin{cases} \dot{x}(t) = Ax(t) + Bu(t) + Df_a(t) + Md(x,u,t) \\ y(t) = Cx(t) \end{cases} \quad (8)$$

We assume that the system (8) satisfies the following conditions:

Assumption 1. The unknown input disturbance $d(x,u,t)$ and the actuator fault f_a satisfy $\|d(x,u,t)\| \le d_0$ and $\|f_a\| \le \alpha_0$, where $d_0 > 0$, $\alpha_0 > 0$ is a known constant.

Assumption 2. The system satisfies that (A,B) is stable and (A,C) is observable.

Assumption 3. There is a positive scalar δ, which satisfies $rank \begin{bmatrix} \delta I_n + A & D \\ C & 0 \end{bmatrix} = n + q$.

The actuator fault is considered as part of augmented state to build the augmented system.

Definition: $\bar{x}(t) = \begin{bmatrix} x(t) \\ f_a(t) \end{bmatrix}$. The following augmented system (9) can be obtained:

$$\underbrace{\begin{bmatrix} I_n & \delta^{-1}D \\ 0 & I_q \end{bmatrix}}_{\bar{E}} \underbrace{\begin{bmatrix} \dot{x}(t) \\ \dot{f}_a(t) \end{bmatrix}}_{\dot{\bar{x}}(t)} = \underbrace{\begin{bmatrix} A & 0 \\ 0 & -\delta I_q \end{bmatrix}}_{\bar{A}} \underbrace{\begin{bmatrix} x(t) \\ f_a(t) \end{bmatrix}}_{\bar{x}(t)} + \underbrace{\begin{bmatrix} B \\ 0_{q \times m} \end{bmatrix}}_{\bar{B}} u(t) + \underbrace{\begin{bmatrix} \delta^{-1}D & M \\ I_q & 0_{q \times h} \end{bmatrix}}_{\bar{D}} \underbrace{\begin{bmatrix} \delta f_a(t) + \dot{f}_a(t) \\ d(x,u,t) \end{bmatrix}}_{\bar{f}(t)}$$

$$y(t) = \underbrace{\begin{bmatrix} C & 0_{p \times q} \end{bmatrix}}_{\bar{C}} \underbrace{\begin{bmatrix} x(t) \\ f_a(t) \end{bmatrix}}_{\bar{x}(t)} \quad (9)$$

Then Equation (9) is transformed into Equation (10):

$$\begin{cases} \bar{E}\dot{\bar{x}}(t) = \bar{A}\bar{x}(t) + \bar{B}u(t) + \bar{D}\bar{f}(t) \\ y(t) = \bar{C}\bar{x}(t) \end{cases} \quad (10)$$

where, $\bar{E} \in R^{n+q}$.

4. Design of the Augmented Sliding Mode Observer

When assumptions 1-3 are satisfied, an augmented sliding mode observer is designed to effectively suppress the effect of actuator faults, uncertainties and external disturbances. The state estimation of augmented system is realized, and the dynamic equations of augmented error system and sliding mode state are obtained. The augmented sliding mode observer (11) is designed as follows:

$$\begin{cases} \overline{E}\dot{\hat{x}} = (\overline{A} - L_p\overline{C})\hat{x}(t) + \overline{B}u(t) + L_py(t) + L_su_s(t) \\ \hat{y}(t) = \overline{C}\hat{x}(t) \end{cases} \quad (11)$$

where, $L_p \in R^{(n+q) \times p}$ is the undetermined gain matrix of the sliding mode observer, and $L_s \in R^{(n+q) \times (q+h)}$ is the sliding mode gain matrix of the sliding mode observer. $u_s \in R^{q+h}$ is a non-continuous sliding mode input term, which eliminates the effect of system actuator faults and uncertainties. It is defined as Equation (12):

$$u_s = \begin{cases} -\rho \frac{He_y}{\|He_y\|}, & e_y \neq 0 \\ 0, & e_y = 0 \end{cases} \quad (12)$$

where, $\rho = \delta\alpha_0 + d_0 + \gamma$, and $\gamma > 0$ is any small positive parameter.

For system (11), the sliding mode gain matrix $L_s = \overline{D}$ is designed. The state error of the system is defined as $e(t) = \hat{x}(t) - x(t)$, the output estimation error is $e_y = \hat{y}(t) - y(t)$, and the Lyapunov matrix P satisfies $\overline{D}^T\left(\overline{E}^{-1}\right)^T P = HC$, where, $H \in R^{(q+h) \times p}$ is a parameter matrix determined by the Lyapunov matrix P.

When Equation (11) is subtracted from Equation (10), the deviation system (13) can be obtained:

$$\overline{E}\dot{e}(t) = (\overline{A} - L_p\overline{C})e(t) + L_su_s - \overline{D}f(t) \quad (13)$$

Since $\overline{E}$ is the nonsingular matrix, there must be the matrix $\overline{E}^{-1}$. The error dynamic model of augmented system (14) can be derived by the left multiplication of Equation (10):

$$\dot{e}(t) = \overline{E}^{-1}(\overline{A} - L_p\overline{C})e(t) + \overline{E}^{-1}L_su_s - \overline{E}^{-1}\overline{D}f(t) \quad (14)$$

Lemma 1: There exists a proportional gain matrix $L_p = SX^{-1}\overline{C}^T$ so that C satisfies the Roulth-Holwitz criterion $\overline{E}^{-1}(\overline{A} - L_p\overline{C})$. X^1 satisfies the Lyapunov equation $-\left(\mu I + \overline{E}^{-1}\overline{A}\right)^T X - X\left(\mu I + \overline{E}^{-1}\overline{A}\right) = -\overline{C}^T\overline{C}$ and $\mu > 0$, which satisfies $Re\left[\lambda_i\left(\overline{E}^{-1}\overline{A}\right)\right] > -\mu$.

The proof is as follows:

Since $Re\left[\lambda_i\left(\overline{E}^{-1}\overline{A}\right)\right] > -\mu$, $\forall i \in \{1, 2, \ldots, n+p\}$ is equivalent to $Re\left[\lambda_i\left(-\left(\mu I + \overline{E}^{-1}\overline{A}\right)\right)\right] < 0, \forall i \in \{1, 2, \ldots, n+p\}$, it can be concluded that for $\forall s \in R^+$, Equations (15) and (16) are valid, as follows:

$$rank\begin{bmatrix} sI_{n+p} - \overline{E}^{-1}\overline{A} \\ \overline{C} \end{bmatrix} = rank\begin{bmatrix} \overline{E}^{-1} & 0 \\ 0 & I_p \end{bmatrix}\begin{bmatrix} s\overline{E} - \overline{A} \\ \overline{C} \end{bmatrix} = rank\begin{bmatrix} s\overline{E} - \overline{A} \\ \overline{C} \end{bmatrix} \quad (15)$$

$$rank[s\overline{E} - \overline{A}] = rank\begin{bmatrix} sI_n - \overline{A} & s\delta^{-1}\overline{D} \\ 0 & (s+\delta)I_q \end{bmatrix} \quad (16)$$

By Equations (15) and (16), Equation (17) is established:

$$rank\begin{bmatrix} sI_{n+p} - \overline{E}^{-1}\overline{A} \\ \overline{C} \end{bmatrix} = rank\begin{bmatrix} s\overline{E} - \overline{A} \\ \overline{C} \end{bmatrix} = rank\begin{bmatrix} sI_n - \overline{A} & s\delta^{-1}\overline{D} \\ 0 & (s+\delta)I_q \\ \overline{C} & 0 \end{bmatrix}$$

$$= \begin{cases} rank\begin{bmatrix} sI_n - \overline{A} \\ \overline{C} \end{bmatrix} + q, s \neq -\delta \\ rank\begin{bmatrix} -\delta I_n - \overline{A} & -\overline{D} \\ \overline{C} & 0 \end{bmatrix}, s = -\delta \end{cases} \quad (17)$$

According to Assumption 3, Equation (18) can be obtained:

$$rank\begin{bmatrix} sI_{n+p} - \overline{E}^{-1}\overline{A} \\ \overline{C} \end{bmatrix} = n + q \quad (18)$$

It can be concluded that $rank\left(\overline{E}^{-1}\overline{A}, \overline{C}\right)$ is observable. There exists a matrix L^*, which makes $\left(-\overline{E}^{-1}\overline{A} - L^*\overline{C}\right)$ stable, that is, satisfying $\left(-\mu I - \overline{E}^{-1}\overline{A} - L^*\overline{C}\right)$ is stable. Further, it can be concluded that $\left(-\mu I - \overline{E}^{-1}\overline{A}, \overline{C}\right)$ is observable.

There exists the matrix $X > 0$ satisfies Equation (19):

$$-\left(\mu I + \overline{E}^{-1}\overline{A}\right)X + X\left(\mu I + \overline{E}^{-1}\overline{A}\right) = -\overline{C}^T\overline{C} \quad (19)$$

By choosing proportional gain matrix $L_p = SX^{-1}\overline{C}^T$, Equation (20) can be obtained equivalently:

$$\left(\mu I + \overline{E}^{-1}(\overline{A} - L_p)\right)^T X + X\left(\mu I + \overline{E}^{-1}(\overline{A} - L_p)\right) = -\overline{C}^T\overline{C} \quad (20)$$

According to Lemma 1, $Re\left[\lambda_i\left(\overline{E}^{-1}(\overline{A} - L_p)\right)\right] < -\mu$, $\forall i \in \{1, 2, \ldots, n+p\}$ is satisfied, that is, $\overline{E}^{-1}(\overline{A} - L_p\overline{C})$ satisfies the Routh Holwitz criterion.

The proof is complete.

L_s, L_p and μ_s are decomposed into the following Equation (21):

$$L_s = \begin{bmatrix} L_{s_1} \\ L_{s_2} \end{bmatrix}, L_p = \begin{bmatrix} L_{p_1} \\ L_{p_2} \end{bmatrix}, u_s = \begin{bmatrix} u_{s_1} \\ u_{s_2} \end{bmatrix} \quad (21)$$

where, $u_{s_1} = [I_{q \times q}\ 0_{q \times}], u_{s_2} = [0_{h \times q}\ I_{h \times h}] u_s$. According to $L_s = \overline{D}$, $L_{s_1} = [\delta^{-1}D\ M]$ can be concluded. By matrix decomposition, the dynamic model of the estimated state system can be obtained from Equation (14), as shown in Equation (22):

$$\dot{\hat{x}}(t) = \overline{A}\hat{x}(t) + \overline{B}u(t) + \delta^{-1}Du_{s_1}(t) + Mu_{s_2}(t) - L_{p_1}\overline{C}e(t) - \delta^{-1}D\dot{\hat{f}}_a(t) \quad (22)$$

In order to avoid system flutters, the continuous function approximation method is used [25], u_s can be approximated by Equation (23) with arbitrary precision, as follows:

$$u_s = -\rho \frac{He_y}{\|He_y\| + \varepsilon} \quad (23)$$

where, ε is a sufficiently small normal number. According to the designed nonlinear sliding mode observer, augmented state $x(t)$ and its estimated value can be obtained. According to the definition of $x(t)$, the estimated value $\hat{x}_p$ of original system state x_p and the estimated value $\hat{f}_a$ of actuator fault f_a can be obtained.

5. Design of Active Fault Tolerant Controller for WECS

The output value Γ^*_{ref} of the WECS can reflect the fault information of the actuator. An active fault-tolerant controller is designed for the actuator fault of the WECS. When the actuator fault occurs, the maximum acquisition of wind energy can be achieved. For the WECS, the expression of sliding mode surface is designed as Equation (24):

$$\sigma = a_1 J_t \Omega_h + a_2 J_t \Gamma_G - J_t \dot{\Omega}_h \tag{24}$$

where, $a_1 = -\frac{1}{T_{sm}}$ is the time constant of sliding mode control convergence speed and satisfies $T_{sm} > 0$. a_2 depends on the steady-state objective of the system, that is, $\lambda_{opt} : \dot{\Omega} = a_1 \Omega_{opt} + a_2 \Gamma_{opt} = 0$, λ_{opt} is the optimum tip speed ratio, Ω_{hopt} not in any equation is the optimum value of high speed shaft speed, Γ_{hopt} not in any equation is the optimum value of generator electromagnetic torque, then $a_2 = -a_1 \frac{\Omega_{opt}}{\Gamma_{opt}}$.

Because the actual control system will be affected by wear and tear, inertia lag of actuator and other factors, the trajectory of the system cannot always be maintained in the switching surface, but switched back and forth around the vicinity, so it is called actual sliding mode dynamics. When the actuator fails, the system can still obtain the desired dynamic characteristics. The general form of sliding mode control law is Equation (25):

$$u = u_{eq} + u_n \tag{25}$$

where u_{eq} is the equivalent control input and u_n is the switching part, as shown in Equation (26):

$$\begin{cases} u_{eq} = \Gamma_G - \frac{T_G}{1+a_2 J_t} \times (a_1 J_t \Omega_h + a_2 J_t \Gamma_G) \times (a_1 - A(\lambda, v)) \\ u_n = -\alpha \text{sgn}_h(\sigma) \end{cases} \tag{26}$$

where, $A(\lambda, v) = \frac{K_v R^2}{i} \cdot \frac{\lambda \cdot C'_p(\lambda) - C_p(\lambda)}{\lambda^2}$, $K = 0.5 \pi \rho R^2$, $C'_p(\lambda)$ is the differential of power coefficient λ and $\text{sgn}_h(\sigma)$ is the hysteresis function with bandwidth h.

When the actuator fault occurs in the WECS, the fault output of the sliding mode controller is as follows:

$$u_1 = u + f_a(t) \tag{27}$$

The control input of the WECS is:

$$\Gamma^*_{ref} = u_1 - \hat{f}_a(t) \tag{28}$$

In this paper, closed-loop feedback control is adopted in the WECS. When the actuator fails, the input signals Ω_h and Γ_G of the controller are changed, which leads to the abnormal control signal Γ^*_{ref} fed back to the system, and then it affects the maximum wind energy capture of the WECS. Through active fault-tolerant control, the fault output is compensated, the output signal of the actuator is corrected, and the active fault-tolerant control target of the actuator fault is realized, so that the performance of the fault system can be restored to the same level as that of the fault-free system.

6. Simulation Analysis

The overall block diagram of active fault-tolerant control for the WECS is shown in Figure 2.

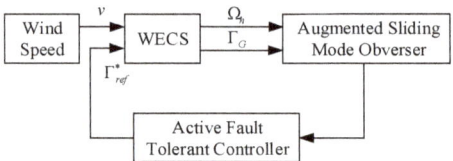

Figure 2. Overall block diagram of fault-tolerant control for the WECS.

The low power and high speed wind energy conversion system based on DFIG is adopted [26]. The simulation parameters are shown in Table 1:

Table 1. Simulation Parameters.

Parameter Names	Parameter Values
Rated voltage V_S	220 V
Rated speed w_S	100 πrad/s
Rated electromagnetic torque Γ_{Gmax}	40 N·m
Electromagnetic time constant T_G	0.02
Air density ρ	1.25 kg/m^3
Transmission efficiency η	95%
Transmission speed ratio i_0	6.25
Blade length R	2.5 m
Moment of interia of high speed axis J_t	0.0092 kg·m^2
Moment of interia of low speed axis J_{wt}	3.6 kg·m^2

At rated wind speed, fixed pitch control is adopted, that is $\beta = 0°$, the wind energy conversion coefficient C_p is determined by the following Equation (29):

$$C_p(\lambda) = -4.54 \times 10^{-7}\lambda^7 + 1.3027 \times 10^{-5}\lambda^6 - 6.5416 \times 10^{-5}\lambda^5 \\ -9.7477 \times 10^{-4}\lambda^4 + 8.1 \times 10^{-3}\lambda^3 - 1.3 \times 10^{-3}\lambda^2 + 6.1 \times 10^{-3}\lambda \quad (29)$$

When the tip speed ratio is $\lambda = 7$, the maximum value is 0.476, which is the best tip speed ratio, where $d(x,u,t) = \begin{bmatrix} \frac{8.5C_p(\lambda)\Omega_h^2}{\lambda^3} & 0 \end{bmatrix}^T$.

It can be seen that Figure 3 is a comparison of the estimated and actual values of the speed of the high-speed shaft speed when the system is fault-free. Figure 4 is a comparison of the estimated and actual values of the electromagnetic torque. As shown in Figures 3 and 4, the sliding mode observer designed in this paper can quickly follow the original state of the system, and the effect of state estimation is satisfactory. The reference value of electromagnetic torque Γ_{ref}^* is shown in Figure 5, and the wind energy conversion coefficient C_p can reach the ideal maximum, which can be kept at about 0.476 in 5~100 s, as shown in Figure 6.

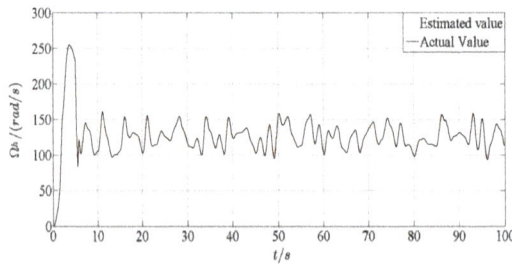

Figure 3. Actual and estimated values of speed Ω_h of high speed shaft.

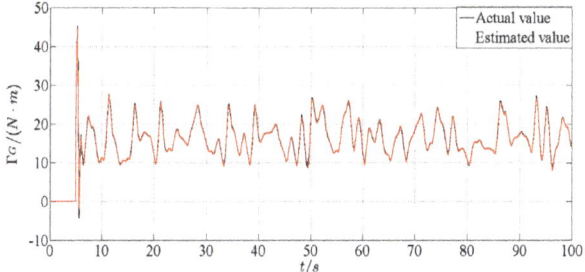

Figure 4. The actual and estimated values of electromagnetic torque Γ_G.

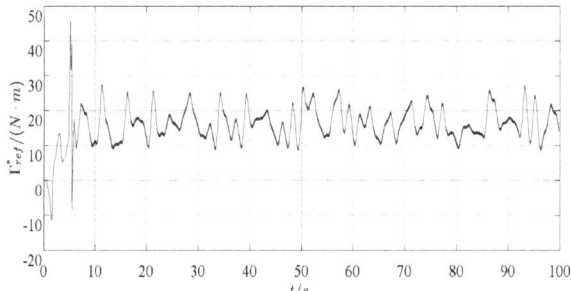

Figure 5. Reference value of electromagnetic torque Γ_{ref}^* without fault.

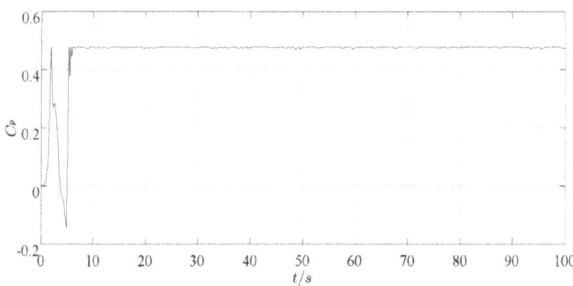

Figure 6. Wind energy conversion coefficient value C_p without fault.

The common actuator faults of the WECS include deviation, drift and so on. Therefore, the simulation design considers the drift fault, deviation fault and the mixed fault function of the actuator, as follows:

$$f_a(t) = \begin{cases} 15 \sin \pi t & 40\text{ s} \leq t < 50\text{ s} \\ 12 & 50\text{ s} \leq t < 55\text{ s} \\ 4.5 + 10 \sin(0.8\pi t) & 55\text{ s} \leq t < 65\text{ s} \end{cases} \quad (30)$$

Figure 7 is a simulation comparison of actuator faults and their reconstructed values. From Figure 7, it can be seen that the augmented observer can accurately reconstruct the actuator fault of the WECS, and directly obtain the fault waveform, magnitude and other information. When the actuator fault occurs, the reference value of electromagnetic torque Γ_{ref}^* changes greatly, as shown in Figure 8. During the period of 40 s~65 s, the wind energy conversion coefficient has seriously deviated from the optimal value and has a large fluctuation range, which cannot be maintained in the optimal position, as shown in Figure 9. By comparing the fault value of performance parameters with that of intact fault-free values, it can be seen that when the actuator fault occurs, the performance of the

WECS is affected to a certain extent, resulting in poor efficiency of wind energy conversion. Active fault-tolerant control can compensate the actuator fault better. The reference value of electromagnetic torque Γ^*_{ref} after fault-tolerant control can approximately follow the actual fault-free state, as shown in Figure 10. When the actuator fault occurs, the WECS can still achieve maximum capture. The tip speed ratio of the system fluctuates near 7 m/s. The fault-tolerant wind energy conversion coefficient is shown in Figure 11.

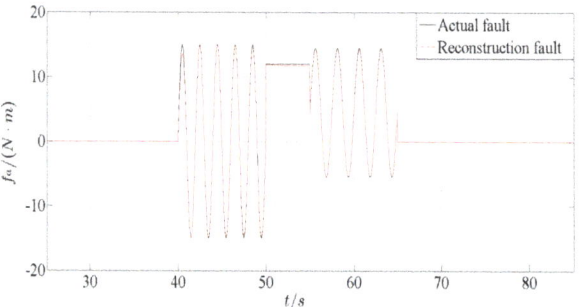

Figure 7. Actual and reconstructed values of actuator fault f_a.

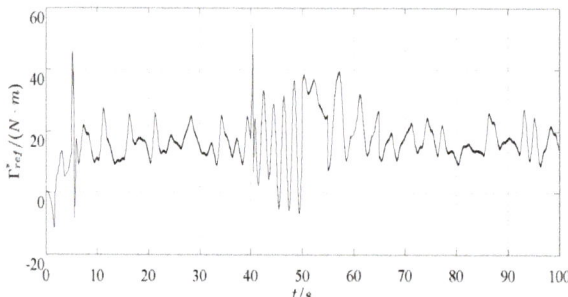

Figure 8. Reference value of electromagnetic torque Γ^*_{ref} when actuator fails.

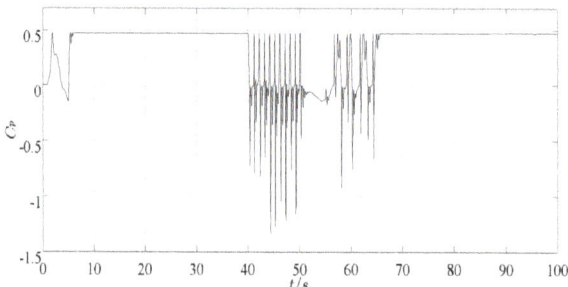

Figure 9. Wind energy conversion coefficient value C_p when actuator fails.

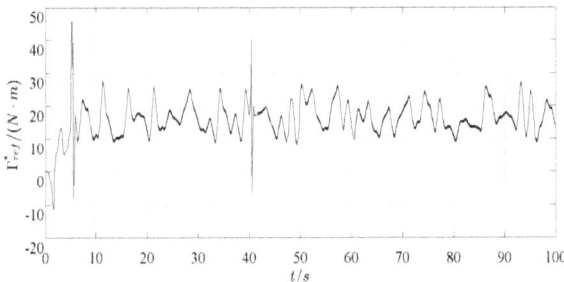

Figure 10. Reference value of electromagnetic torque Γ_{ref}^* after fault-tolerant control.

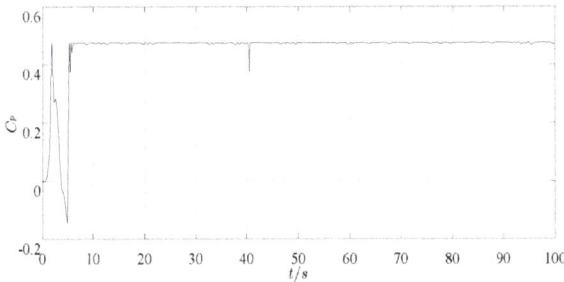

Figure 11. Wind Energy Conversion Coefficient Value C_p without Fault after Fault Tolerant Control.

7. Conclusions

In this paper, the problem of state estimation and fault reconstruction for uncertain WECS are discussed when the actuator fault occurs. An augmented sliding mode observer is constructed by splitting the non-linear term of the state equation of the WECS into uncertain parts of the system and augmenting the state. Then the robust fault reconfiguration observer is designed, and the equivalent output method is used to reconstruct the actuator fault, which has strong robustness. The active fault-tolerant controller designed ensures the stable input of the system and captures the maximum wind energy.

Author Contributions: X.W. conceived the experiment and wrote the paper; Y.S. helped in the experiment and writing.

Funding: This work was supported by the National Nature Science Foundation under Grant 61573167.

Conflicts of Interest: The authors declare no conflict of interest.

References

1. Attya, A.B.; Dominguez-Garcia, J.L.; Anaya-Lara, O. A review on frequency support provision by wind power plants: Current and future challenges. *Renew. Sustain. Energy Rev.* **2018**, *34*, 483–490. [CrossRef]
2. Yang, Z.; Chai, Y. A survey of fault diagnosis for onshore grid-connected converter in wind energy conversion systems. *Renew. Sustain. Energy Rev.* **2016**, *66*, 345–359. [CrossRef]
3. Hang, J. An Overview of Condition Monitoring and Fault Diagnostic for Wind Energy Conversion System. *Trans. China Electrotech. Soc.* **2013**, *28*, 261–271.
4. Kamal, E.; Aitouche, A. Robust fault tolerant control of DFIG wind energy systems with unknown inputs. *Renew. Energy* **2013**, *56*, 2–15. [CrossRef]
5. Vidal, Y.; Christian, T.; José, R.; Acho, L. Fault Diagnosis and Fault-Tolerant Control of Wind Turbines via a Discrete Time Controller with a Disturbance Compensator. *Energies* **2015**, *8*, 4300–4316. [CrossRef]

6. Kamal, E.; Aitouche, A.; Ghorbani, R.; Bayart, M. Robust fuzzy fault-tolerant control of wind energy conversion systems subject to sensor faults. *IEEE Trans. Sustain. Energy* **2012**, *3*, 231–241. [CrossRef]
7. Badihi, H.; Zhang, Y.; Hong, H. Wind Turbine Fault Diagnosis and Fault-Tolerant Torque Load Control Against Actuator Faults. *IEEE Trans. Control Syst. Technol.* **2015**, *23*, 1351–1372. [CrossRef]
8. Wu, D.; Jin, S.; Shen, Y.; Ji, Z. Active fault-tolerant linear parameter varying control for the pitch actuator of wind turbines. *Nonlinear Dyn.* **2017**, *87*, 475–487. [CrossRef]
9. Wu, Z.Q.; Yang, Y.; Xu, C.H. Adaptive fault diagnosis and active tolerant control for wind energy conversion system. *Int. J. Control Autom. Syst.* **2015**, *13*, 120–125. [CrossRef]
10. Benbouzid, M.E.H.; Beltran, B.; Amirat, Y.; Yao, G.; Han, J.; Mangel, H. Second-order sliding mode control for DFIG-based wind turbines fault ride-through capability enhancement. *ISA Trans.* **2014**, *53*, 827–833. [CrossRef]
11. Mekri, F.; Elghali, S.B.; Benbouzid, M.E.H. Fault-Tolerant Control Performance Comparison of Three- and Five-Phase PMSG for Marine Current Turbine Applications. *IEEE Trans. Sustain. Energy* **2013**, *4*, 425–433. [CrossRef]
12. Beltran, B.; Benbouzid, M.E.H.; Ahmed-Ali, T. Second-order sliding mode control of a doubly fed induction generator driven wind turbine. *IEEE Trans. Energy Convers.* **2012**, *27*, 261–269. [CrossRef]
13. Benelghali, S.; Benbouzid, M.; Charpentier, J.F.; Ahed-Ali, T.; Munteanu, I. Experimental Validation of a Marine Current Turbine Simulator: Application to a Permanent Magnet Synchronous Generator-Based System Second-Order Sliding Mode Control. *IEEE Trans. Ind. Electron.* **2011**, *58*, 118–126. [CrossRef]
14. Ha, Q.P.; Trinh, H. State and input simultaneous estimation for a class of nonlinear systems. *Automatica* **2004**, *83*, 1779–1785. [CrossRef]
15. Lee, D.J.; Park, Y.J.; Youn, S. Robust H∞ sliding mode descriptor observer for fault and output disturbance estimation of uncertain systems. *IEEE Trans. Autom. Control* **2012**, *57*, 2928–2934. [CrossRef]
16. Zhang, J.; Shi, P.; Lin, W. Extended sliding mode observer based control for Markovian jump linear systems with disturbances. *Automatica* **2016**, *70*, 140–147. [CrossRef]
17. Gao, Z.; Liu, X.; Chen, M.Z.Q. Unknown Input Observer-Based Robust Fault Estimation for Systems Corrupted by Partially Decoupled Disturbances. *IEEE Trans. Ind. Electron.* **2016**, *63*, 2537–2547. [CrossRef]
18. Alipouri, Y.; Poshtan, J.; Zarch, M.G. Generalized Sliding Mode with Integrator Controller Design Using a Discrete Linear Model. *Proc. Inst. Mech. Eng. Part I J. Syst. Control Eng.* **2014**, *228*, 677–689. [CrossRef]
19. Zhang, K.; Jiang, B. Fault Diagnosis Observer-based Output Feedback Fault Tolerant Control Design. *Acta Autom. Sin.* **2010**, *36*, 274–281. [CrossRef]
20. Liu, X.; Gao, Z.; Chen, M. Takagi-Sugeno Fuzzy Model Based Fault Estimation and Signal Compensation with Application to Wind Turbines. *IEEE Trans. Ind. Electron.* **2017**. [CrossRef]
21. Carroll, J.; Mcdonald, A.; Mcmillan, D. Reliability Comparison of Wind Turbines with DFIG and PMG Drive Trains. *IEEE Trans. Energy Convers.* **2015**, *30*, 663–670. [CrossRef]
22. Qu, Y.B.; Song, H.H. Energy-based coordinated control of wind energy conversion system with DFIG. *Int. J. Control* **2011**, *84*, 2035–2045. [CrossRef]
23. Kenyon, M. Energy-Reliability Optimization of Wind Energy Conversion Systems by Sliding Mode Control. *IEEE Trans. Energy Convers.* **2008**, *23*, 975–985.
24. Zhang, Z.; Verma, A.; Kusiak, A. Fault Analysis and Condition Monitoring of the Wind Turbine Gearbox. *IEEE Trans. Energy Convers.* **2012**, *27*, 526–535. [CrossRef]
25. Shen, Q.; Jiang, B.; Cocquempot, V. Adaptive Fuzzy Observer-Based Active Fault-Tolerant Dynamic Surface Control for a Class of Nonlinear Systems with Actuator Faults. *IEEE Trans. Fuzzy Syst.* **2014**, *22*, 338–349. [CrossRef]
26. Munteanu, I.; Bratcu, A.I.; Cutululis, N.-A.; Ceanga, E. *Optimal Control of Wind Energy Systems*; Springer: London, UK, 2008.

© 2019 by the authors. Licensee MDPI, Basel, Switzerland. This article is an open access article distributed under the terms and conditions of the Creative Commons Attribution (CC BY) license (http://creativecommons.org/licenses/by/4.0/).

MDPI
St. Alban-Anlage 66
4052 Basel
Switzerland
Tel. +41 61 683 77 34
Fax +41 61 302 89 18
www.mdpi.com

Energies Editorial Office
E-mail: energies@mdpi.com
www.mdpi.com/journal/energies

www.ingramcontent.com/pod-product-compliance
Lightning Source LLC
LaVergne TN
LVHW071952080526
838202LV00064B/6728